"Wes Chaar brings to the table a formidable trove of more than two decades of experience in the realm of data science, an odyssey that stretches across an impressively eclectic array of disciplines.

His recent book, *Data Independence*, not only encapsulates this vast experience but also positions him as a wise observer of technological evolution. His book opens a gateway to new dialogues about data—a field that, as Wes convincingly argues, should be accessible and inviting to all. In reading his reflections, one is struck by the portrait of a man not merely present at the crossroads of history but actively shaping the confluence of technology and daily life across multiple industries. Wes Chaar invites us all to step through the door he holds open, into a conversation about data that is as expansive as it is essential."

—JOE RAAEN
Data Employee

"Wes Chaar has consistently been at the forefront of innovation, always pushing the envelope with unique ideas and solutions. Renowned as a disruptor, he has often introduced concepts that challenge the status quo and encourage a fresh perspective on prevailing issues.

In *Data Independence*, Wes continues on this path by tackling the issue of data privacy. Drawing from over two decades of experience in data science, he demystifies the complex dynamics of the data economy, revealing how our personal data is used and monetized. But more importantly, he proposes a groundbreaking framework for reform—a systematic approach centered on data consent, data control, and data currency. *Data Independence* isn't just about understanding the mechanics of data privacy; it's a call to action. It encourages us to think differently about our offline and online presence and to take decisive steps toward reclaiming our data autonomy.

For anyone interested in the future of privacy and the role of data in our society, Wes Chaar's book is an essential read. It promises not just to inform but to transform how we interact with our digital world. As always, Wes remains a steadfast disruptor, charting the path toward a better future and world."

—SUZANNE MORRIS
Telecom Executive

"During a time when data is gathered from all kinds of sources, distributed through many channels, and leveraged to enhance business and marketing operations, consumers should question their rights and abilities to choose what happens to data specifically related to them, their activities, and their lives.

Data serves as the quiet, invisible, yet highly informative digital DNA, revealing more about us than we might realize. As such, we should consider the management of this information in the same way we'd protect our biological DNA from those able to leverage our personal fabric purely for company profit.

Younger generations, digital natives, have grown up with the notion of cameras watching their every move in retail environments and in public, devices capturing features such as eyes, facial bone structure, thumbprints, voice signatures, and other data gathering as a part of their daily lives. However, privacy and freedom are interconnected constructs when it comes to personal data. Without the choice to research and decide the right paths and purposes for personal data, there is no real data independence.

Dr. Chaar's frameworks for how to consider the privacy-freedom paradigm will surely appeal to the common consumer and open their eyes to the tragedy of a corporate status quo involving data capture without real choice.

He illustrates how the release of this personal digital DNA provides limited benefit to the consumer, and even sometimes risks or negative consequences, with mostly positive outcomes to companies.

The straightforward yet brilliant ideas put forth provide new ways for consumers to encourage policymakers to allow for data independence and may change the way you see the world, your interactions with brands, and your connection with businesses as part of our highly technology-driven lives.

It is my hope this work of art and science will inspire consumers and motivate governmental institutions to rethink our way of operating in a more data independent environment for the future."

—BEVERLY WRIGHT
Data Science and AI Thought Leader

"After dedicating over two decades to the data and analytics industry, Wes Chaar recognized that the systems he helped develop were becoming increasingly problematic, sparking concerns about their long-term implications. In his book, he offers a historical analysis of how well-intentioned engineers over the past forty years have built a vast market that now drives the global economy but often at the cost of personal privacy. He discusses the concept of privacy as a fundamental, albeit implicitly recognized, right by the founding fathers of the United States. The book serves as an essential guide to understanding the journey of personal data and illustrates, through straightforward examples, how something as simple as casual conversations can quickly translate into targeted advertisements. While acknowledging the widespread consensus on the issues of data privacy, Wes proposes a novel approach to empower individuals with control over their data, suggesting a framework no more invasive than the financial oversight seen on Wall Street."

—NIKOLAOS VASILOGLOU
AI Industry Expert

DATA INDEPENDENCE

DATA INDEPENDENCE

WES CHAAR, PH.D.

Reclaiming Privacy
in an Era of
Evolving Tech

Published by Advantage, Charleston, South Carolina.
Member of Advantage Media.

ADVANTAGE is a registered trademark, and the Advantage colophon is a trademark of Advantage Media Group, Inc.

Printed in the United States of America.

10 9 8 7 6 5 4 3 2 1

ISBN: 978-1-64225-975-9 (Hardcover)
ISBN: 978-1-64225-974-2 (eBook)

Library of Congress Control Number: 2024912160

Cover design by Matthew Morse.
Layout design by Lance Buckley.

This publication is designed to provide accurate and authoritative information in regard to the subject matter covered. It is sold with the understanding that the publisher is not engaged in rendering legal, accounting, or other professional services. If legal advice or other expert assistance is required, the services of a competent professional person should be sought.

Advantage Media helps busy entrepreneurs, CEOs, and leaders write and publish a book to grow their business and become the authority in their field. Advantage authors comprise an exclusive community of industry professionals, idea-makers, and thought leaders. Do you have a book idea or manuscript for consideration? We would love to hear from you at **AdvantageMedia.com**.

In memory of my father, Samir Chaar.

CONTENTS

INTRODUCTION

What digital DNA do you trail behind you as you live your life?

The 1997 science fiction movie *Gattaca* depicts a dystopian future in which the DNA of each individual is collected, tracked, and categorized. The government monitors this DNA in a digital bank of "biometrics" that decides what job you can have, how much money you can earn, and what role and status in society you will play. *Gattaca* shows how much DNA we shed on a daily basis. Those who wish to evade detection by the government in this future world collect their hair follicles, nail trimmings, and skin cells to ensure they leave every room they depart as clean as before they arrived. A particularly harrowing scene involves Ethan Hawke rushing back to his office chair to tweeze a single hair follicle off of his keyboard.

While the creators of *Gattaca* portend that their movie depicts a "not-too-distant" future, they did get one thing about our twenty-first-century world right: Every person leaves a trail that can be used to track them. Not just DNA, but our digital DNA.

In each of our daily interactions, from the coffee we buy every morning to the streaming services we watch at night, the book we bought, the webpage we visited, the trip we are planning or took, our daily commute and errands, the post we made online, etc., data is created that links us to our activity in the world. Much like DNA, those who know how to collect and track this data can use it to create a clear picture of who you are and how you move through the world. They even use this information to predict your behavior and to influence your decisions. Unlike your DNA, however, much of this data is completely invisible to you, meaning you won't be able to grab that stray hair of your data to prevent its collection.

Today, all of these disparate pieces of an individual's information can be "stitched together" (something we will cover later in the book) to supercharge our human capacity for solving business problems. While personal data integration offers significant promise for businesses, the broader landscape of data utilization encompasses much more, extending its transformative power across various sectors.

For example, meteorological data, which has seen huge advancements in the past years, can now be used to pinpoint major storms with incredible accuracy. In medicine, analyzing health data offers the ability to give advanced diagnoses and forecast health issues far beyond what was once thought possible, and it's saving lives every day. Organizations and businesses across the globe use data to hone in on inefficiencies and bottlenecks to ensure smooth daily operations. There is even a whole industry of data experts—me included—that assists these institutions with their data and helps them make the most out of it. With all of these innovations, it's hard to imagine how such a useful tool could be wielded against us and our freedoms.

While data is, indeed, a powerful tool, those in the upper echelon of the tech industry and business world wield it freely against everyday folks in their pursuit of more money and even more power. For example, media conglomerates collect and analyze data created every day to serve up advertisement after advertisement to people who fit the perfect "customer criteria." And some, if not most, of this data borders on the personal. Beyond media conglomerates, other industries also harness data for diverse purposes: retail companies use purchasing data to tailor their marketing strategies and product offerings, financial institutions leverage consumer data to assess credit risk and personalize banking services, hedge funds analyze various data assets to determine where they will invest, and many more. For each of these sectors, the line between beneficial data use and privacy infringement is increasingly blurred.

But it isn't just big corporations that use data to monitor an individual's activity.

Governmental organizations from around the globe are also collecting and buying data to monitor, locate, and track the average citizen. In the United States, a company called X-Mode quietly bought geolocation data harvested from dating apps and sold it to the Department of Homeland Security and the Internal Revenue Service.[1] The fact that governmental organizations use the path of purchasing commercial data feeds demonstrates the surveillance net that digital commerce has created is feasible to monitor, track, collect intelligence, and more.

Called a "legal gray area," the National Security Agency (NSA) has been buying *Americans'* internet activity from data brokers.[2] Who needs

---

1   Lee Fang, "IRS, Department of Homeland Security contracted firm that sells location data harvested from dating apps," The Intercept, 2022, https://theintercept.com/2022/02/18/location-data-tracking-irs-dhs-digital-envoy/.

2   Charlie Savage, "NSA Buys American's internet data without warrants, letter says," *New York Times*, January 25, 2024, https://www.nytimes.com/2024/01/25/us/politics/nsa-internet-privacy-warrant.html.

warrants when they can buy our digital DNA? And you can be certain if we are doing it to our citizens, other countries are too.[3]

Forget *Gattaca*. The dystopia lives today with our freedoms being trampled on in the name of convenience and commerce!

As a data science expert with over twenty years of experience working with and understanding the implications of data, I've served in various contributor and leadership roles in the data, analytics, data science, machine learning (ML), and artificial intelligence (AI) space for many companies and have contributed to data practices in the airlines, hospitality, e-commerce, advertising, media, marketing, consumer-packaged goods (CPG), retail, and data-as-a-service (DaaS) industries.

Yet, here I am—sounding an alarm. When it comes to the issues surrounding data privacy, I've been at the front lines of many of these concerns. In my career, I have written algorithms, devised code, and consulted with companies to best use data to forward their product innovation, bottom line, and other corporate goals. And what I see today is beyond disquieting. It is both frightening and infuriating.

My path to this point had some twists and turns. My doctorate is actually in science and mathematics. When I was young, I was obsessed by the airline and aerospace industries. I'd devour any book or movie that was flight- or space-related, from airplanes, astronauts, and space shuttles to satellites. I spent hours in my room envisioning trips to other planets and galaxies. In my young mind, the path to those faraway places would start with becoming a pilot. There was only one "small" problem: my parents would not let me take flight lessons and instead kept me bound to my home planet.

---

3   Steven Feldstein, "Governments Are Using Spyware on Citizens. Can They Be Stopped?" 2021, https://carnegieendowment.org/2021/07/21/governments-are-using-spyware-on-citizens.-can-they-be-stopped-pub-85019.

I needed a new plan. In college, I pursued degrees in computer and communications engineering. From there, I moved on to a master's in computer science with a focus on systems and control engineering. This is where I became interested in the ways things or systems work together. After all that schooling, I decided to continue my studies and return to my first love—I earned my doctorate in aerospace engineering.

Doctoral candidates in aerospace engineering generally study aerodynamics, fluid mechanics, material science, guidance navigation control (GNC), or some other specialty. The Ph.D. candidates that focus on GNC develop the mathematics and the algorithms that make it possible for spacecraft to navigate off of our planet and to other planets in and beyond our solar system. They devise algorithms to plot safe travel through the stars, almost like they're piloting the ships themselves. As a Ph.D. student, GNC was the field I was drawn to because I was always interested in navigation, hence the desire to be a pilot. In a sense, it didn't matter if you were behind a computer or on the flight deck because, with GNC, it's all navigation, right?

During my Ph.D. studies, I went on to work at the University of Texas Center for Space Research. There, I developed algorithms for NASA's Jet Propulsion Laboratory (JPL) to allow spacecraft navigation for deep space missions, meaning travel even farther into space, and with little to no human intervention. In 1997, I introduced the use of AI, ML, and genetic algorithm (GA) to navigate in space. My approach centered on developing a framework for machines and algorithms to learn from each other and create new and more powerful algorithms than the initial population of algorithmic solutions.

That is, mathematics solutions learn from each other in order to create new mathematics that would be more accurate for space travel. This allowed for new navigation systems to be created autonomously.

In 1997, my work was tested in parallel to standard approaches used for the NASA JPL Mars Pathfinder Mission.

But while I was learning, developing all these algorithms, and using them to travel between the stars, I was still interested with the airline industry. I'd monitor the news for stories related to advancement in air travel technology, or when a new plane was commissioned. It was through my fascination with this industry that I learned about their approach to data.

Way back in the late 1980s, before Big Data was a familiar concept to other industries, airline companies were using data to do customer segmentation, selling, and allocating specific inventory to the "right customer at the right time"—creating a whole new field of mathematical modeling called revenue management (RM). Think of it simply like this: Every airline knows a certain number of seats sold will actually fly empty—people get sick, change plans at the last minute, and so on. Furthermore, there are different segments of consumers (e.g., business travelers, vacationers, and leisure travelers) and at different price points. Airlines had this and much more information figured out from data. They began to develop predictive and prescriptive models that would help them manage their inventory— the seats on the plane. When I learned about this, I thought about the algorithms I was using in space travel and how they could benefit airlines back on Earth. That's how I started my data science career at Sabre Holdings and then at Delta Air Lines, where I brought my new mathematical modeling approaches to that industry.

From there, my career in data exploded. My time with the airline industry led to more opportunities in tech sectors and other business verticals like advertising, media, CPG, retail, and more. With the airline industry, I was looking at how they could maximize revenue from various aggregations of air travel consumers, meaning segments.

In various other verticals, I had to learn how to connect what seems initially unrelated data to form a picture of an individual (within permissible data uses by law). My team and I took massive datasets to use identity graphs based on inputs from a customer relationship management (CRM) database, connected devices across the country, and other datasets. We used this information to understand consumers at every single point, learning how to present them with the right message at the right time. This technology worked to serve these consumers content that would generate the right feeling, the right promotion, for whatever my client's objectives were. Consumers just like you, reading this.

If it sounds like Big Brother—it is.

The scary thing was, it worked! And—even eerier—some of the early, well-intended, and innocent breakthroughs in these data milestones had occurred around 1984. I felt like someone who had uncovered a terrible and powerful Orwellian secret.

> Winston kept his back turned to the telescreen.
> It was safer, though, as he well knew, even a
> back can be revealing. —GEORGE ORWELL, 1984[4]

While working for various companies, I remember looking at the algorithms with my team and saying, "Wow, humans are so predictable!" Based on the work I did there, and the datasets that were available to me, I could foresee their behavior with high accuracy. Later, this became about predicting *sentiment*. I could tell exactly what someone might buy when they're getting groceries, what kind of streaming media they might consume, what type of content will resonate best with an individual to attain a particular outcome (e.g., a sales lift, a conversion goal), and much more.

---

4   George Orwell, *1984* (New York, NY: Signet Classics, 1961).

These companies and other companies worked within the confines of the laws, surely. And yes, they are companies looking to succeed, be profitable, and provide services that are attractive to people. However, what happened is that we, the individuals, lost our *freedom* in the process. I started to question and didn't like what I observed. In fact, it made me viscerally uncomfortable to have this much insight into hashed/masked identities of people I didn't know. Which is why I am writing this book—to start a conversation and to propose a transparent and respectful solution around data.

Privacy is freedom.

We all deserve to live our lives without feeling like we're being monitored. Despite the benefits that come from data, it has created a world where our data—our digital DNA—is being collected and, more importantly, *exploited* in ways that most people will find shocking. In this book, I will take you through some of the most common techniques that companies and entities use to track, predict your behavior through data, maximize their profits, gain market share, etc.

I will also provide a new paradigm for data *consent,* data *control,* and data *currency*—we will delve into the "C3s" later. A new paradigm for data is needed to ensure the freedom of every individual in this country and the world. I will outline a new approach to data that will ensure the power of data is controlled by the people, the rightful owners of the data. In the future, your data could be accessed through a Personal Data Vault Key (PDVK) in which you—and only you—have the key. This powerful aggregation of data could then be used by organizations that you choose who will outline exactly what they're doing with your data, assuming you decide to participate in the data economy. In this new data ecosystem, the people have the power, and their control of their privacy is enshrined as sacred. I will explain how this concept is a solution to most of the data privacy issues we have today—and will explain how in part II.

This book will not only outline the history of data, and the current and future issues, but I will also present a "Data Constitution." We the people need to gain control of our digital DNA. And those changes start with you. But in order to even begin the conversation, that Data Constitution will set out some fundamental principles we all should and can agree on—just like the Founding Fathers did as America was being born.

Everyday people—not just those in Big Tech and Big Business—need to know the intricacies of our current data landscape and how they can advocate for themselves, their privacy, and gain complete control of their personal data. Change comes slowly. But if we start talking about the data topics of ownership, consent, control, and currency today, there's no telling what our future data ecosystem will look like. With your help, we can give the power of data to the people and make a more equitable and efficient future for us all. With people in control of their data and its usage, they can participate in the data currency, they have the "data vote" on which new technologies they support, especially what AI should and should not be allowed to do, and much more.

So, join me as we take back our data.

PART I

Our Data
Rights

CHAPTER 1

What Is Privacy, and How Have We Lost Our Way?

Privacy means people know what they're signing up for, in plain language, and repeatedly. I believe people are smart. Some people want to share more than other people do. Ask them.

—STEVE JOBS

In 2012, Target, the Big Box Store with everything, sent coupons for diapers and cribs to a young girl in Minneapolis. However, this young girl's parents had yet to discover that she was pregnant. The company crawled through her recent purchases, flagged her as high on a "pregnancy score," and sent her coupons designed for new mothers.[5] To better secure a sale, Target tracked this young woman's intimate purchases, violating her privacy.

In 2013, Yahoo! experienced a data breach that compromised *over three billion* account holders' data.[6] The company hid this from the public for over three years. In 2021, a hacker posted LinkedIn data

---

5   Keith Wagstaff, "How Target knew a high school girl was pregnant before her parents did," *Time*, 2012, https://techland.time.com/2012/02/17/how-target-knew-a-high-school-girl-was-pregnant-before-her-parents/.

6   Michael Hill and Dan Swinhoe, "The biggest data breaches of the 21st century," November 8, 2022, accessed October 20, 2023, https://www.csoonline.com/article/534628/the-biggest-data-breaches-of-the-21st-century.html.

from 700 million account users to the Dark Web.[7] Equifax was hacked in 2017 and 147 million consumers' information was compromised.

Even more frightening, 23andMe confirmed that personal information for 6.9 million customers was breached. They also discovered that some of the breach intentionally targeted Chinese and Ashkenazi Jews.[8]

Daily, we are bombarded with news about our data privacy being compromised—often stolen from household-name companies. But there are countless ways our privacy is breached.

Data is also viewed as currency—it has value in and of itself. For example, when American Airlines needed collateral for a secured government loan, their Loyalty Rewards program (aka data) was valued at between $19.5 and $31.5 billion.[9] This was solely for data—not planes, not personnel, not terminals and hangars. Our data, our buying habits, our privacy, can be sold to the highest bidder.

In a Securities and Exchange Commission (SEC) filing in 2023,[10] Delta revealed that data generated $5.5 billion from American Express—their biggest contributor to profits. Delta added 8.5 million SkyMiles members (and their data!) in 2022. They added 1.2 million cobranded Amex cardmembers.[11] And it all added up to billions of dollars.

---

7   "Biggest data breaches in history," https://www.consumernotice.org/data-protection/breaches/biggest-in-history/.

8   Rebecca Carballo, Emily Schmall, and Remy Tumin Jan, "23andMe breach targeted Jewish and Chinese customers, lawsuit says," *New York Times*, 2024, https://www.nytimes.com/2024/01/26/business/23andme-hack-data.html.

9   Adam Levine-Weinberg, "Is American Airlines' loyalty program really worth $31.5 billion?" The Motley Fool, June 17, 2020, accessed October 20, 2023, https://www.fool.com/investing/2020/06/17/is-american-airlines-loyalty-program-really-worth.aspx.

10 "SEC Filings Details," 2023, https://ir.delta.com/financials/sec-filings/sec-filings-details/default.aspx?FilingId=16314380.

11 Gary Leff, "3 Facts Delta just dropped about SkyMiles and their American Express partnership," 2023, https://viewfromthewing.com/3-facts-delta-just-dropped-about-skymiles-and-their-american-express-partnership/.

When we log onto our social media networks, or onto various shopping sites, or our health care portal, or any of the countless technological interactions and tools and apps we use each day, we often are shown an updated privacy policy (in fine print). For most of us, it's an inconvenient pop-up. We scroll down quickly and hit "accept," agreeing to incursions on our date privacy without really being sure of what it is we're agreeing to.

Even when we don't agree—when we forbid the use of cookies and other data-tracking tools—the average person assumes that act of clicking "don't accept" is enough, when, in fact, blocking cookies does virtually nothing to ensure your privacy. What I know from my years of experience in this world is that we are all—each and every one of us—being tracked with methods that invade our privacy in ways most of us cannot even begin to realize (though this book will hopefully make this invasion of privacy apparent!).

But with all this talk of privacy, and before we move on in this book, we should come to a baseline understanding of what privacy is, why we claim to value it—and how it has eroded as we moved more and more of our existence into an online space.

THE PHILOSOPHY AND FRAMEWORK OF PRIVACY

Why do we even care about privacy?

If you think of movies like Florian Henckel von Donnersmarck's Oscar-winning masterpiece *The Lives of Others,* or other films like *Minority Report, Gattaca,* or *Enemy of the State,* they depict surveillance states, and movie watchers are understandably appalled by the tactics we see up on the screen. We are appalled precisely because we are conditioned to believe in a fundamental right to privacy—that goes back centuries for Europe and the United States. Other countries may condition their citizens to accept the trade-offs of a surveillance state (such as, perhaps, lower crime rates) in exchange for loss of privacy.

For most of us, though, privacy is a sacred concept.

We might trace some of the early thoughts on privacy to the English philosopher John Locke in his work "Two Treatises of Government."[12] Locke proposed that individuals have a natural right to "life, liberty, and property." In this context, privacy is seen as an extension of these natural rights, enabling people to maintain autonomy over their lives and belongings.

In the United States, the founding fathers did not explicitly mention the right to privacy in the Constitution. However, the Supreme Court has interpreted the Bill of Rights to provide a constitutional right to privacy, derived from the first, third, fourth, fifth, and fourteenth amendments.

In the fourteenth amendment, citizens of the United States have the right to "life, liberty, or property" that cannot be infringed on by the government without "due process of law." This reading, combined with other amendments' interpretations, were part of the *Griswold v. Connecticut (1965)* decision, where the US Supreme Court weighed in on the matter of privacy.[13] This landmark case centered on the freedom of individuals to use contraception without the interference of the government. In the *Griswold* decision, the Supreme Court acknowledged that there were "zones of privacy" in which the government could not interfere. One of the biggest of these "zones" was the right to marital privacy, but this decision has also been used to ensure other intimate family matters. While

---

12   John Locke, "Two Treatises of Government (or Two Treatises of Government: In the Former, The False Principles, and Foundation of Sir Robert Filmer, and His Followers, Are Detected and Over," 1689.

13   https://supreme.justia.com/cases/federal/us/381/479/#:~:text=Connecticut %2C%20381%20U.S.%20479%20(1965)&text=A%20right%20to%20privacy%20 can,contraception%20by%20married%20couples%20illegal.

this might (and recently, rumblings in politics say perhaps not) protect us from government overreach, it says nothing of *companies*.

American philosopher and jurist Louis Brandeis, in his seminal article "The Right to Privacy," cowritten with Samuel Warren in 1890, argued that privacy is "the right to be let alone."[14] This right, according to Brandeis, is essential to human dignity and personal development. Brandeis's perspective on privacy has had a significant impact on privacy jurisprudence in the United States.

Are we being "let alone" when it comes to our data?

Or are we being fed, like the humans in the pods in *The Matrix?* Fed data while, in turn, our data becomes the fuel for the "machines"? Or are we like marionettes on a string, manipulated by the puppeteers who control our data? (See figure 1.1.)

With data collection and monitoring rising across the country, we must understand what our rights are as citizens and advocate for ourselves. In this context and as it relates to data, we can look to the fourth mendment, which protects us from unreasonable searches and seizures. Some legal scholars have used this argument to extend our government's notion of privacy to our data (a position I heartily concur with). Legal scholar Morgan Cloud posits the following:

> The broad concept of property can apply to digital information because that theory protects more than tangible things. As understood during the century leading up to the Founding, the broad concept of property included a person's rights, ideas, beliefs, and the creative products of her labor.[15]

---

14   Samuel D. Warren II and Louis Brandeis, "The right to privacy," Harvard Law Review, 1890.

Figure 1.1: Who or what is controlling us?

However, the founding fathers and all of the legal scholars of the 1700s and 1800s, into the 1900s, could not have foreseen where we are now. (I do like to think, though, that if Thomas Jefferson materialized in front of us right now, he would concur that our personal data falls into the realm of privacy.)

DATA AND PRIVACY

In chapter 3, we will discuss the "data creep" of how little by little, we have given over our data—and it has been *taken* until we now find ourselves in a privacy quandary. Ownership, for example, is an interesting concept for something as ethereal as data.

Consider this: Who owns our data? Is it ours (individually)? Or does it belong to the company that harvests it from our everyday interactions?

This is at the crux of our data and privacy conundrum. I believe that the full data picture belongs to us, as individuals, as it is our personal behavioral DNA. A company conducting commerce surely needs records of the commerce or transactions between the individual and entity. For example, my daughters like to shop in certain clothing stores, a friend of mine's husband has a favorite tech store where he buys all their gadgets, and a sneaker company that offers a free pair after purchasing four needs to track how many sneakers a consumer has bought. Those companies knowing what their consumers have purchased in the past can help guide their marketing, discounts, promotions, etc. However, I firmly believe the transactions should not be used to stitch with other data sources from other entities in order to form a 360-degree view of the individual.

---

15   Morgan Cloud, "Property is Privacy: Locke and Brandeis in the Twenty-First Century," *American Criminal Law Review* 55, no. 1 (2018), accessed October 25, 2023, https://www.law.georgetown.edu/american-criminal-law-review/in-print/volume-55-issue-1-winter-2018/property-is-privacy-locke-and-brandeis-in-the-twenty-first-century/#:~:text=The%20broad%20concept%20of%20property,creative%20products%20of%20her%20labor.

I believe we need a framework that is in the control of the individual. *We the people* should decide, via a PDVK/solution that I am proposing in this book, if we wish to participate, how to participate, what to achieve or gain, and more, when it comes to sharing our data. It is for the individuals to invite, if they wish, companies and entities into their homes and lives by sharing their data. And, it is for the individuals to change their decision at any point in time and disinvite entities they previously invited to use their data.

AUTONOMY AND INDIVIDUALITY

Privacy is essential for fostering personal autonomy and individuality. It allows us to develop our thoughts, emotions, and personal relationships without undue interference or scrutiny from others. We recognize the concept of "our home is our castle," and most believe what we do in the privacy of our homes or our bedrooms is not the business of our government—and we would likely add it is not the business of corporate entities either. However, I would posit that our internal thoughts are the ultimate "castle," and our thoughts and behavioral DNA should be ours to share—or not.

In addition, as the writers of our Constitution predicted, a free society needs free discourse among its citizens. With today's technology, those conversations and debates are no longer happening in the pubs and taverns of the founders, but online, on messaging boards, and social platforms where folks share and comment on posts from their social circles. While it may seem obvious that we can't be targeted for this discourse, the massive amounts of data we create online has drawn the attention of those who wish to control us.

This goes far beyond concepts like "cancel culture" or even censorship. For example, already, some military contractors are quietly buying data that allows them to track specific groups. An app called *Muslim Pro*, which has over ninety-eight million users, helps followers of Islam track and guide their daily prayers. Unfortunately, that convenient app sold geolocation data to a third-party broker X-Mode, which then sold that crucial data to military contractors that work with the Department of Defense and other governmental agencies.[16] How can we use these digital spaces to voice our opinion and connect with others who share that opinion if it leads to a violation of our privacy?

FREEDOM OF ASSOCIATION AND ASSEMBLY

Privacy is essential for preserving freedom of association and assembly. In a democratic society, individuals must be able to join together to pursue common goals, express dissenting opinions, and participate in political activities without fear of retaliation. Privacy safeguards this right by ensuring that people can associate with others and form groups without being subject to government surveillance or public scrutiny.

Again, how is our data being used to track those we gather with online? The importance of privacy in protecting the freedom of association was demonstrated in the landmark Supreme Court case *NAACP v. Alabama (1958)*.[17] The court held that the NAACP had a constitutional right to protect its membership lists from disclosure, recognizing that privacy was crucial in allowing the organization to pursue its civil rights objectives without undue interference.

---

16   Joseph Cox, "How the US military buys location data from ordinary apps," *Vice*, November 16, 2020, accessed October 28, 2023, https://www.vice.com/en/article/jgqm5x/us-military-location-data-xmode-locate-x.

17   "NAACP v. Alabama," https://firstamendment.mtsu.edu/article/naacp-v-alabama-1958/.

PROTECTION FROM DISCRIMINATION AND PROFILING

Privacy is vital in protecting individuals from discrimination and profiling based on their personal information. In a world where vast amounts of data are collected and analyzed, individuals face the risk of being unfairly targeted or stigmatized based on their race, religion, sexual orientation, or other personal characteristics. Privacy protections ensure that individuals can maintain control over their personal information, reducing the likelihood of discriminatory profiling or targeting.

For instance, the European Union's General Data Protection Regulation (GDPR) enshrines strong privacy rights for individuals, including the right to access, correct, and delete their personal data, and the right to object to profiling based on their data. These privacy protections aim to safeguard individuals from discrimination and profiling in the digital age. (But, as I will also explore in this book, legislation is not a panacea and rarely works the way it is intended.)

PRIVACY AS A CHECK ON GOVERNMENT POWER

Privacy serves as a vital check on government power by limiting the scope of state surveillance and interference in individuals' lives. Unchecked government surveillance can lead to the abuse of power, suppression of dissent, and erosion of democratic values. Privacy protections ensure that government agencies are subject to oversight and accountability, helping to preserve individual freedoms and maintain the balance of power in a democratic society.

Edward Snowden's revelations about the NSA's mass surveillance programs brought to light the potential dangers of unchecked government surveillance, sparking a global debate about privacy rights and state power. In the wake of these disclosures, legal and policy reforms have been implemented to strengthen privacy protections and rein in government surveillance, reinforcing the connection between privacy and the preservation of individual freedom. However, the availability of (under today's laws) legal commercial capabilities that are indeed a surveillance system for commerce means these laws are easily bypassed, and any entity can pay for and attain the desired data.

I believe the solution I put forth in this book could solve many of these issues.

MILESTONES IN DATA PRIVACY

These privacy issues, especially as related to our data, have been developing for as long as computers have existed. However, some digital natives may not even realize this timeline because data and computers have been part of their lives since birth. Here is a brief look at where we started, where we are now—and where this Brave New World may be headed (we will cover the future innovations of Big Tech and data, including AI, in chapter 10).

Centralized Databases and Mainframes (1960s through 1970s)

These were the days when databases and mainframes were massive, filling rooms—before personal computers, and long before we held all of our personal lives on a smartphone device we fit in our pocket. However, in terms of data privacy, this marked the nascent beginnings of centralized data storage and collecting and storing of personal information (which could be retrieved by those with permission to access this data).

Credit Reporting and Databases (1970s through 1980s and Beyond)

Credit reporting agencies started collecting data and maintaining centralized financial histories. However, according to Brookings and others, the error rate in credit reports is almost 75 percent![18] Yet as we all know, this (often incorrect) data can impact our lives on everything from car loans and mortgages to job screenings—and they collect a tremendous amount of information about us from medical debt to when we are late to pay a credit card.

Loyalty Programs (1980s and Beyond)

This is where I begin to enter the story with industries I joined (in the late 1990s) in my own career. We will discuss this more in the next chapter, but airlines (frequent flier miles), grocery stores (loyalty cards), and other industries began collecting mass amounts of data on individual purchases. That data provided companies with incredible insights into buying habits and preferences—and later, algorithms could offer predictive consumer modeling.

Personal Computers and the Internet (1980s and 1990s)

You may be old enough to remember the painful days of dial-up internet connectivity. However, this was the era that ushered the computer into our *homes,* which increased the risk for data breaches and unauthorized access. It also ushered in an era of people creating online lives—even if in those early days it was internet chat rooms and bulletin boards.

---

18   Aaron Klein, "The real problem with credit reports is the astounding number of errors," September 28, 2017, accessed October 28, 2023, https://www.brookings.edu/articles/the-real-problem-with-credit-reports-is-the-astounding-number-of-errors/.

Online Advertising and the Era of Cookies (1990s through 2000s)

As the internet evolved and online retail and commercialization came into play, online tracking started with cookies. This enabled targeted advertising and much more. While we are all so used to shopping online today, it can sound hard to believe that Amazon only started online in 1995, and a store like the GAP only offered online shopping beginning in 1997. Today its brick-and-mortar stores are becoming increasingly rare. And that is true for many retail stores. (Since 2017, over 1,200 malls have shuttered.[19])

Social Media (2000s through Today)

Our lives have very much shifted over the last two decades. So much of it is online—and one of the reasons is social media and the platforms important to so many. Meta, X, Instagram, TikTok, Snapchat, WhatsApp, and others collect massive amounts of data on us—everything from our relationships to our preferences and behaviors. Algorithms "feed" users content that reflects those preferences, leading to significant privacy concerns. Add the fact that 90 percent of kids ages thirteen to seventeen are on social media and these privacy issues are heightened (with added concerns of safety).[20]

---

19 Zohar Gilad, "Will e-commerce be the savior of brick-and-mortar stores?" *Forbes*, https://www.forbes.com/sites/forbestechcouncil/2024/02/12/will-e-commerce-be-the-savior-of-brick-and-mortar-stores/?sh=2fdc9e493c2a.

20 "Social media and teens," American Academy of Child and Adolescent Psychiatry, March 2018, no. 100, accessed October 15, 2023, https://www.aacap.org/AACAP/Families_and_Youth/Facts_for_Families/FFF-Guide/Social-Media-and-Teens-100.aspx.

Smartphone Location Data (2000s on)

The handy devices we carry in our pockets and purses can track our physical location and data (as can SmartTags, FitBits, and other smart watches, and our cars' GPS). This just adds another layer to our data. Now it is easy to tell where we are in the world. It's also easy for *others* to know where we are in the world.

Internet of Things (2010s on)

The Internet of Things (IoT) ushered in additional smart devices. Everything from our thermostats to our refrigerators can now collect data on us, our household, and our behaviors. Door cams, the Echo Dot, Google Home, and smart glasses are but a few more IoT devices—and they are becoming more sophisticated all the time.

Big Data, Machine Learning, and Artificial Intelligence (2010s on)

Now we are in new territory. Companies can use Big Data, ML, and AI to discern deeper insights about people (even things they may not be aware of themselves).

I want to pause here to note that I am not against the proliferation of data analytics and technology. I've made data my career, after all. It's about *consent*, *control*, and *currency*—our three Cs.

With all of these innovations, it's hard to imagine how such a useful tool could be wielded against us and our freedoms. But it is—every day.

CHAPTER RECAP

Now we have a background in privacy—how it impacts much more than what goes on in our homes or in our online life. It affects our freedoms on a fundamental level—something recognized for centuries. Even if the thinkers and philosophers of years past could not anticipate the world we live in now, the principles and concepts are the same.

Later, I will provide a new paradigm for data consent, data control, and data currency. A new paradigm for data is needed to ensure the freedom of every individual in this country and the world. I will outline a new approach to data that will ensure the power of data is controlled by the people, the rightful owners of the data. In this new data ecosystem, the people have the power, and their control of their privacy is enshrined as sacred. But to get there, we need changes—and that will not happen overnight, especially with Big Tech fighting such changes and exploiting loopholes in any legislation.

However, we need to start this conversation.

Not only do we need to start it, but we also need to do so with transparency.

Privacy is an integral aspect of freedom, enabling individuals to maintain autonomy, express their thoughts and opinions, and pursue their personal goals without undue interference or scrutiny. As digital technologies continue to evolve, the relationship between privacy and freedom becomes increasingly complex, with new challenges and opportunities emerging.

To ensure that privacy remains a cornerstone of individual freedom, it is imperative that data control is given to the rightful owners of the data: the individual.

To devise a transparent solution, let's start, as the saying goes, by "following the money." As we follow the money at stake and what allows the realization of this money, we land on the driver of the digital economy: data. The data is, in the first place, the outcome of everyone's personal and private activities. The data is part of our individual DNA, our purchase DNA, our behavioral DNA, our interests DNA, our habits DNA, etc. By embracing a fundamental approach to privacy that empowers the individual to control their own data and vote its usage, we can preserve the essential connection between privacy and freedom, fostering a just and flourishing society for generations to come.

When I write "follow the money," we should be very transparent about the value we are discussing. A report by McKinsey & Company published in January 2022 estimated that companies could unlock trillions in value annually by fully harnessing the potential of data and analytics. The report notes that many organizations are not fully leveraging their data assets, and that the potential value could be unlocked through better data management, analytics, and decision-making.[21]

I am not suggesting that the digital age does not provide conveniences that many people like. I am suggesting transparency between businesses and individuals. I am proposing control back to the individual of their complete data elements and picture. I am asking businesses to compete for these complete pictures of data. I want the individual to exercise their "data vote."

Now, before we move on to our next topic, the next chapter is going to introduce you to five consumers—five people just like you and me, our neighbors and friends. Each of them demonstrates a way in which data is accessed by companies and weaponized, in some ways, against unsuspecting digital citizens.

---

21   McKinsey Digital, "The data-driven enterprise of 2025," January 2022, accessed December 10, 2023, file:///C:/Users/erica/Downloads/the-data-driven-enterprise-of-2025-final.pdf.

CHAPTER 2

Our Consumer Profiles

We cannot be mere consumers of good governance, we must be participants; we must be co-creators.

—ROHINI NILEKANI

In order to better understand just how much of our lives and data are tracked, we will visit with the following consumers throughout this book, who are meant to be composites of typical people going about their lives. Typical people—whose lives and data are being used by corporations, likely without the consumers being aware of it.

Erica, a mom of four, wakes up before dawn so she can enjoy a cup of tea in peace before the chaos of a typical family morning. She watches her Wi-Fi-enabled television, clicking through her cable guide, and stops at CNN (perhaps we know her politics now), then moves on to *Bluey.* It is now assumed she either has a preschooler, or she is too exhausted to change the channel.

Wi-Fi-enabled TVs have Automated Content Recognition technologies, which capture what and when consumers are watching. It is another digital channel by which a consumer or individual can be tracked and reached. In turn, this is connected to a larger ID Graph of every person.

An important concept used by commercial companies and discussed in this book is this ID Graph, an intersection between an individual's various reward programs, laptop, mobile device (iPhone,

Android, etc.), Wi-Fi-enabled TV, satellite radio, social channels, etc. It allows the identification of the individual, tracking the person's digital activities and enables the messaging to the individual via various channels. Activities are analyzed and used by various entities to gain insights, thus enabling the messaging, engagement, and selling of various products and services to an individual.

While she waits for her kids, she pops Eggos in the toaster for the toddler (the teenagers tend to forage, she thinks). Those waffles? She used a coupon from her loyalty card at her favorite grocery store. Speaking of which—when she drove there two days ago, a satellite took a picture of the parking lot for the store.

Her loyalty card captures her shopping transactions. Her consumer purchasing behavior is enriched with various other data sources so that the company offering the rewards program knows much more about her—and us all. For example, what are the details of the household associated with the rewards program? Details include number of people who live in the household, income, gender, children/no children, pets, etc. Additional data sources can be lifestyle interests, geo data, and more. A concatenation of data is created aiming for a 360-degree view of every individual.

And that satellite picture, what was that about? A hedge fund on Wall Street has capabilities to analyze the volume of cars (against other years and time periods) to assess the health of the retailer's business. The firm buys this data from the satellite provider.[22]

She gets everyone off to school (not an easy feat), driving a Toyota minivan that Big Data knows she bought by the relentless searches she did when car shopping. Because her geo data is

---

22   Laura Counts, "How hedge funds use satellite images to beat Wall Street—and Main Street," Haas News, Berkeley Haas, May 28, 2019, accessed February 6, 2024 https://newsroom.haas.berkeley.edu/how-hedge-funds-use-satellite-images-to-beat-wall-street-and-main-street/.

known, she is made an offer on her way back home to her office to increase her propensity to stop at Starbucks and make a purchase. Predictably (for Big Data), she stops for that Starbucks (now we know she likes caramel macchiatos with soy milk—oh, and she gets soy milk coupons with her loyalty card). Next she goes into her home office. Before she checks her emails for work, she orders something online at Bloomingdale's.

Her shopping at Bloomingdale's, along with other online searches and household information, is used to understand her state of mind regarding purchases—what she is likely to purchase and at what price points. She usually buys expensive work attire—but always at end-of-season sales.

Every step of her digital interaction is tracked, mined, and monetized continuously and relentlessly.

And most likely, she has no idea.

Jonathan is an executive—a travel warrior. Some days, he wakes up in a hotel and forgets what city he is in. He rises one morning and books his travel to Vancouver for a conference (now we know he travels business class and likes to stay at the Four Seasons). However, before he left for Vancouver, he was thinking about purchasing a Mercedes and fiddled around on his laptop searching for a model he liked (did he really need a sedan, or could he be sportier?). Mostly, he *hovered* and did not click. However, on the plane to Vancouver, he is bombarded with display ads for the Mercedes he most coveted.

Airlines providing Wi-Fi on the plane is the norm today—typically free of charge. Free? Not really—there is no such thing as a free lunch! It is "free" because the airline decided at some point it is more valuable to offer free Wi-Fi and capture the passengers' digital data while using its Wi-Fi. The data's monetization value outweighs the cost of Wi-Fi.

Jonathan's travel journeys are understood in detail. His travel provides monetization opportunities to the airline, credit card company, car service, advertising agency, and other entities. Jonathan has an ID Graph, and every digital interaction he makes is captured. His booking of the flight provides an opportunity to the airline to sell the information to car service companies offering him rides from home to the airport, from his airport arrival destination to the hotel, etc. Hotels, too, buy this information so as to sell the right product to the right passenger at the right price.

The exact seating for every passenger is known. By using the plane's Wi-Fi and entertainment screen, each passenger is targeted with the most relevant products and services for them. The seating of each passenger is known and hence the specifics, peculiarities, and individuality of each traveler is known. In other words, what's their profile? What are their interests? What did they search for recently? Do they have a new higher-paying job and more disposable income? Everything about everyone is known. This connected screen to each passenger is an example of connected TV, meaning Wi-Fi-enabled TV connected to an individual or household. Knowing with certainty who is watching the screen allows the personalization of messages and offers, hence increasing the propensity of achieving a certain outcome.

Rose is eighty-three and really wants to stay independent in her own home as she ages. But her craftsman bungalow in Seattle is sorely in need of a roof. She searches for roof tiles and looks online. Based on her other purchases, Big Data knows she is elderly. Her data is sold to a scammer who sells her a roof for three times the going rate (and does a shoddy job, no less), but as a target, she also catches the attention of real estate companies and financial institutions. These entities are monitoring her data to gauge her health status and predict when her house might hit the market. Worse, after she searched for some personal medical devices and considered a scooter to get around, she is slammed with ads and calls from medical device companies. Furthermore, financial institutions are analyzing her transactions and online behavior, contemplating the right moment to offer her a home equity loan, reverse mortgage, or other financial services, leveraging the data to understand her potential needs.

Kara is a thirteen-year-old girl whose parents have recently allowed her more freedom on the internet. She has a TikTok, a Snapchat, an Instagram profile, *and even a "Finsta"* (which Mom and Dad don't know about). She is obsessed with K-pop, Timothee Chalamet, all things soccer, and she has started following many makeup and diet gurus. Consequently, a large portion of what Kara sees in her feed are reminders of body imperfection and the need for makeup, supplements, and trendy fast fashion. Kara, of course, has no idea how much filters and lighting, and even AI models, are impacting her views of what "beauty" is. She is "fresh meat" to the digital world, and her digital path started, probably, when her mother searched for baby clothes online.

But now, Kara is her own digital consumer, and her online activities start weaving a complex web of data that will be of interest to a multitude of industries. Retail and CPG companies will analyze her preferences to predict and influence future purchases. Media companies will tailor content to her evolving tastes. Insurance companies might use her data to infer lifestyle choices. Beauty brands will track her interest in makeup and skincare products. Airlines and hotels will see potential in her as she grows into a traveler. Banking and financial services will gauge her financial behavior as she becomes an adult. Additionally, educational technology, health and wellness, automotive, entertainment, telecommunications, and real estate industries will all have a stake in her long-term customer value, each seeking to understand and leverage her preferences and behavior over time.

Thus, Kara's journey through the digital realm is not just a passage of personal growth; it's a valuable narrative in the eyes of a data-driven market, continually shaping her as a customer across multiple stages of her life.

Tom, a thirty-something accounting executive, finds himself at a crossroads, grappling with career stagnation and the emotional aftermath of a recent breakup. His online activities, a mosaic of searches and digital interactions, paint a vivid picture of his current state. He scours the internet for advice on coping with breakups, reads articles on overcoming heartache, and frequently visits forums discussing emotional healing. These activities, combined with his increased engagement with content related to depression and mental well-being, attract the notice of numerous companies.

He starts noticing an influx of targeted ads: St. John's wort, an over-the-counter herbal remedy often touted for alleviating depression symptoms; a plethora of mental health apps offering support and counseling; and even dating sites, suggesting that perhaps loneliness is his underlying issue. His social media feeds, influenced by his recent search history, begin to fill with articles on self-improvement, travel destinations for solo travelers, and promotions for singles events and local meetups. Additionally, lifestyle and wellness brands begin targeting him with ads for gym memberships and yoga classes, assuming a potential interest in physical health as a way to cope.

All these targeted suggestions, stemming from Tom's online behavior, further blur the lines between helpful guidance and overwhelming intrusion. What was initially a quest for personal answers and solace in the digital space becomes a bombardment of options and paths, leaving Tom more uncertain than comforted.

CHAPTER RECAP

Now that you have met our profiles, we will revisit them from time to time. They are you—and me. Let us now discuss these digital breadcrumbs and data creep a little more fully, as we explore in chapter 3.

Following the Digital Breadcrumbs—and the Money

> *Relying on the government to protect your privacy is like asking a Peeping Tom to install your window blinds.*
>
> —JOHN PERRY BARLOW, CYBERLIBERTARIAN, POLITICAL ACTIVIST, AND FOUNDING MEMBER OF THE ELECTRONIC FRONTIER FOUNDATION

In chapter 1, I shared a timeline of how digital tracking and our data have been collected from the nascent beginnings of mainframes to AI and today. In this chapter, I will share more details about how our data is collected—and what it is worth.

First, let's understand, in a big-picture sense, what sort of digital transformation has happened and how it changes the world.

Let's start with advertising. Once upon a time, there were limited mediums to reach the people advertisers wanted to target. Forgetting direct mail for a moment, the avenues were magazines, newspapers, television, or radio. However, advertisers had to take an "educated guess" as to who the person was that was seeing their ad. They guessed age and gender of the person or household who viewed a program by the type of show being aired. If it was Monday Night Football (again, before more modern views on gender and interests), the guess was that the audience was mostly male—the better audience to advertise Gillette razors to. If the show was a Saturday morning cartoon, the audience was children (even if plenty of grown-ups loved *Bugs Bunny*).

Ads in "women's magazines" like *Good Housekeeping* were aimed at stay-at-home moms (termed "housewives," in an antiquated word).

Advertisers and media had, essentially, these buckets of how they approached people by each demographic. This is how Nielsen started (the company that determined television ratings and what was a hit or not). But with the digital transformation, advertisers now know who that person or viewer is. They are not just in a certain bucket, they are multidimensional. It has become a one-on-one; it has become personalized.

Not only do they know who you are, but they also know how to reach you. And they know—in terms of *sentiment*—what you respond to.

Digital transformation, the internet explosion of data, and now countless, targeted channels allow the individual to be reached in a new multitude of ways. But, more importantly, it's personalized. Digital is one-on-one, whereas the previous way of communication was mass communication.

They know who you are.

Turning our attention to the banking industry, the digital revolution has reshaped the way we manage our finances. In the past, banking was predominantly a physical experience. Customers visited brick-and-mortar branches for transactions, consultations, and financial advice. Personal banking was broadly based on customer segments, with banks offering standard products like savings accounts, mortgages, and loans without deep personalization. They relied on demographic data and general financial behaviors to categorize customers into broad groups like "young professionals," "families," or "retirees."

With the advent of digital banking, the industry has moved beyond these broad categories to a more nuanced, individual-focused approach. Banks now harness extensive data from online transactions, app usage, and digital interactions to gain deep insights into individual financial habits and preferences. This shift enables them to offer highly personalized services, from tailored investment advice to customized loan offers. Digital banking platforms use algorithms to predict customer needs, offer real-time financial advice, and even anticipate future life events that could impact financial decisions.

They know who you are.

Similarly, the travel industry has undergone a remarkable transformation in the digital era. Gone are the days when travel planning was a generic experience, often involving visits to travel agents who offered standard package deals with little customization. Travelers were grouped into basic categories like "families," "business travelers," or "backpackers," with offerings broadly targeted to these segments. The industry operated on assumptions about traveler preferences based on these limited categories.

The digital transformation has enabled a seismic shift toward personalized travel experiences. Travel companies now have access to a wealth of data about individual travelers' preferences, past trips, and even dream destinations gathered through online searches, bookings, and reviews. This data empowers travel agencies, airlines, and hotels to tailor their offerings to each traveler's unique preferences. From personalized travel recommendations to customized itineraries, the industry can now cater to the specific needs and desires of each traveler. The rise of AI and ML further enhances this personalization, allowing for dynamic pricing, real-time travel updates, and even predictive suggestions for future trips.

They know who you are.

In the realm of the insurance industry, the digital shift has brought about a fundamental change in how insurance is understood, sold, and managed. Traditionally, insurance relied on broad demographic data and statistical probabilities to assess risk and determine premiums. Customers were often grouped into large categories based on age, location, and basic lifestyle factors. This approach, while effective in some ways, often overlooked the nuances of individual risk profiles and needs.

With the digital era, however, the insurance industry has seen a dramatic shift toward personalization and precision. Today, insurers utilize a vast array of data points, from online behavior to IoT device data like fitness trackers and connected home devices. This allows them to construct a detailed and dynamic risk profile for each individual, leading to more accurate premium setting and personalized coverage options. For instance, a customer's driving behavior captured through a telematics device can directly influence their car insurance rates. Health insurers might use data from fitness trackers to offer wellness incentives. This level of personalization not only benefits the insurers in terms of risk management but also empowers customers with more tailored, relevant insurance products and potentially lower costs.

They know who you are.

In today's digital ecosystem, virtually every industry—from technology, health care, and finance to education, entertainment, and transportation, and more—possesses the capability to construct a 360-degree view of individuals. Leveraging advanced data analytics, AI, and the vast swaths of digital footprints we leave online, these sectors can now assemble detailed profiles that encompass not just demographic details but also personal preferences, behaviors, and even predictive future actions. This profound insight means that whether you're browsing a website, using a mobile app, or interacting with a smart device, industries have the tools to know who you are, tailor their offerings to your specific needs, and anticipate your future requirements with unprecedented precision.

They know who you are.

STITCHING A PROFILE

We will only cover a couple of technological capabilities used in the construction of a full profile. This section and the book are not meant to be a comprehensive technology review of all the techniques used in the domain. This section demonstrates just a couple of the methods— and as technology and AI advances, these tools will only evolve.

Your ID Graph

When you (the individual whose data is being exploited) started becoming personalized, your privacy was invaded because fundamentally the world of data now knows exactly who you are because you have an identity graph (ID Graph). The digital world knows where you are at any point in time, in anything that you do.

Let's borrow Amazon Web Services' definition of ID Graphs (because they would know!):

"An identity graph provides a single unified view of customers and prospects based on their interactions with a product or website across a set of devices and identifiers. An identity graph is used for real-time personalization and advertising targeting for millions of users. This is done by linking multiple types of identifiers to form a consistent, unified view of the customer. An identity graph can also store profile data and easily connect new consumer identifiers to profiles.

"Identity graphs can provide a 360° view of customers to understand the customer journey in chronological order or make recommendations to close a deal. An identity graph also helps you build customer data platform (CDP) solutions with an emphasis on privacy regulation compliance. Identity graphs are a key solution for many advertising technology and marketing technology companies, as well as brand and marketing organizations, advertising agencies, holding companies, and web analytics providers."[23]

If you recall our profiles from the last chapter, you may have been skeptical—how could they know all that about me, about these characters?

The ID Graph is (one of) the how.

But what about privacy laws? As we will discuss in part II of this book, laws—in the world of Big Data—are meant to be gotten around. They have largely been ineffective—and I do not anticipate that changing. In the world of Big Data, loopholes were meant to be found and taken advantage of.

One of those loopholes is the "clean room." (Just *one* of the loopholes—there are many!)

The Clean Room

A clean room is anything but, in my personal opinion. While we will dive into this in more detail later, for now understand that a "clean room" is a place where two different entities can share and "stitch" data—and it is all perfectly legal. Each entity gains data belonging to the other through this process. They do share data, but in a regulatory space that supposedly does not violate consumers' privacy (wink-wink, nudge-nudge).

Look at it like this: Imagine an entity possessed information about Erica, our mother in our individual profile. The *only* information this entity knows is that she is a mom of more than two kids, she is a New Yorker, and she frequently flies for work as a journalist. Now imagine another entity knows her book-buying history, and also

---

23   Amazon Web Services, "What is an identity graph?" https://aws.amazon.com/ neptune/identity-graphs-on-aws/#:~:text=What%20is%20an%20identity%20 graph,targeting%20for%20millions%20of%20users.

what political organizations she donates to. But neither entity knows what the other does. These entities both enter a clean room, through a process that masks her personally identifiable information (PII) to create a combined picture of the two data assets. The "data" of what they each have about this individual is stitched together. They each leave knowing the "full(er) picture." That full picture is then attached or used with Erica's ID Graph.

The person who is the subject of this clean room exercise has no idea this is taking place.

Advertising Example: Personalization Empowered through the ID Graph and Clean Room

It used to be that marketing and advertising were very much "spray and pray." Send out ads at certain demographics—lots of ads, repetitive ads—and hope something stuck. But I can't single out marketing and advertising. They only do this at the behest of the banking, retail, travel, insurance, grocery, etc., industries.

Yet we are all so very different. Even if I am a fifty-something man, I may be very different from someone down the block who is also my age. You can make some assumptions about us (perhaps we both golf or enjoy single malt scotch). But you could only guess traits about us based on our ages, and where we lived, or what television shows people in our demographic often watch.

"Creative," a term in the industry for the message or specific advertisement used, was therefore often broad.

Today, the creative has become more personalized; it is specific to every person. By knowing much more about me—or you—we can get digital ads on our phone, our smart TVs, our airline seat TVs, or on our laptops as we browse that are more aligned with what we supposedly want to see. That is made possible through personalization. And that is possible through algorithms. (Some of which I played a part in creating—but now I see we have lost our privacy, and our *control* over our privacy.)

Thus, for every consumer, digital advertisers can now deliver something different for a multitude of contexts because they have a much clearer, perhaps even complete, picture of the individual. What's the propensity to buy this or to do this action? Once again, we are the marionettes, and "they" are the puppeteers.

In a simple example, think of an ad for beer in three different places and contexts. The first ad is set in Minnesota in January and shows ice fishing with cold beer being kept on literal lake ice. All the people in the ad are middle-aged men. The second is the *same beer,* but it is shown to people contemplating heading to Florida for a winter getaway and its ad is full of tropical colors, sand, turquoise water, flip-flops, and Jimmy Buffet music. Finally, the third ad is sent to those whose data show they are college students, and it depicts a cool nightclub and a bunch of twenty-somethings leaving, climbing the stairs up to the roof, and starting their own party under the stars— with their beers, of course.

Three different contexts. Same product.

WHAT HAPPENED TO FREEDOM?

I've said it before. Privacy is freedom. Let me say it again for those in the back (and I will repeat it throughout the book): *Privacy is freedom.*

We have already discussed why the concept of privacy is so important. Now we are aware there has been such data creep that we have lost our data privacy little by little. (For example, if you are middle-aged or older, when you signed up for Facebook way back, you probably NEVER thought you would be sharing everything from recipes to politics to your kids' prom pictures.)

But this loss of privacy has been twofold. We all like to believe that what we say in the privacy of our homes, what we talk about, what we watch on television or our iPads or phones, what we search for online, … should be our personal "castle." But they are spying on us.

In fact, algorithms are making our choices close to *deterministic.*

According to *Webster's Dictionary*, "determinism" is "a theory of doctrine that acts of the will, occurrences in nature, or social or psychological phenomena are casually determined by preceding events or natural laws."[24]

In other words, as I mentioned in the beginning of this book, humans are very "predictable." We can write predictive algorithms . . . and know what those marionettes are going to do as we pull the strings.

The problem with freedom is that you have to be able to understand this element to people's privacy—this digital element. The people in data *understand* this (and work to mine it), and the average individual does not. Again, many people tend to think if they check the box that says, as an example, "no" to cookies, their privacy must be safe.

The Data Invasion

The data creep was perhaps inevitable. But with personalization, it went from "creeping in" to a downright invasion. Big Data invaded our privacy.

Think about it. Who asked our digital overlords to know every single detail about us and personalize us? I did not—and I am sure you did not either. I also certainly did not ask any entities or companies, advertisers, apps, or social media feeds to get into my thoughts, feelings, and whatever else is swirling around inside me in order to make money off of me.

I didn't ask to be a marionette.

---

24 *Merriam-Webster*, "Determinism," https://www.merriam-webster.com/dictionary/determinism.

Mercenary Data Creep

Remember our profile Jonathan, the travel warrior? Because all of his travel is booked online, Big Data knows a *lot* about him. They know where he is going next month (London, then off to Munich, and finally Copenhagen). They have some inkling (better than an inkling) of how much he makes because they are aware of every restaurant he has booked via online reservations in each city. He favors Michelin-starred places. On each plane, they know which fare class he has booked (always business class, more indication of his income or status in his company).

As he settles into his comfortable seat and reclines, advertising comes on the screen—aligned with his recent search for an S-Class sedan. His phone aligns with the same advertising because Jonathan used Wi-Fi at the airport and at the beginning of his flight, while everyone was boarding and before he turned his phone onto airplane mode. Meanwhile, he thinks he got the Wi-Fi for free because he is such a prolific flier and has logged more than 190,000 miles this past year.

Oh, our poor, naïve Jonathan. It is *not* because they love him but because they are making more money. This is connected TV. This is the cream of the crop of advertising. Because Jonathan is sitting in seat 2B in business class, he cannot go anywhere. He's *a captive audience*.

On the plane, he cannot go to his kitchen, or out to eat; he cannot flip the remote control.

Guess how the airlines, advertisers on behalf of companies, or any "you name it" company make even more money?

They can now relay Jonathan's information to a ride-share company so they can target him and others when they land in London. Guess what? The trips from the airport to the city and the city to the airport are some of the most profitable routes for ride-share corporations.

In some ways, it is much like a rat in a maze. Advertisers, on behalf of companies, have a very pungent Stilton, and they are luring you, steering you, where they want you to be in the maze (see figure 3.1).

They can even measure attention. How focused are you on what is being presented to you on screens? Some devices have technology that knows if you saw something by tracking your eyes. Understanding what's effective on any person allows the creation of the appropriate message and content to "influence" (let's call it what it really is: manipulate) a certain outcome. For example, Kantar has integrated Realeyes' Attention and facial coding technology into its Context Lab solution for social video and mobile ad measurement.[25] It just emphasizes how we, the masses, are shackled in the digital age—it's *1984* via gentle and what I am terming *soft totalitarianism,* manipulation at scale every second of the day.

The Origins of Data Creep

AIRLINE INDUSTRY EXAMPLE

To be honest, this all started a little more innocently. As you may recall, I mentioned that airlines were inventing these RM systems years ago. It actually all began when President Jimmy Carter was in office. In 1978, he deregulated the airline industry (Airline Deregulation Act).[26]

When this happened, airlines began to formulate their plans: Where should they fly? How much should they price their tickets? How do they estimate demand? This is when the smart people in the

---

25 M. Kalehoff, "Kantar integrates Realeyes attention tech to advance ad effectiveness," https://blog.realeyesit.com/kantar-integrates-realeyes-attention-tech-to-advance-ad-effectiveness?utm_campaign=2023%2011%20-%20Kantar%20x%20Realeyes&utm_content=273343790&utm_medium=social&utm_source=linkedin&hss_channel=lis-PgnsTKKB1V.

26 "S.2493 - Airline Deregulation Act," https://www.congress.gov/bill/95th-congress/senate-bill/2493.

Figure 3.1: Do we have data independence and privacy, or are we rats in a maze?

airlines figured out there was a great deal of available data (in the back office, on mainframes) about who flew certain flights.

Let us just randomly take an 8:00 a.m. flight from Dallas to New York. The airline has the data about all the flights that flew this route at this time in the past. They also know, because flights are planned in the future, how many people are booking for the flight in a typical week. The data also reveals how many tickets are usually booked three weeks out. Then two weeks. The airlines thus have an idea of how the demand builds up the closer it gets to the flight.

This was, essentially, the first aspect of predicting airline demand that was figured out.

Next, they examined the data that demonstrated not everyone who books shows up for the flight. This is called "no-show forecasting." The airlines created this form of forecasting—and let's say for our purposes that our 8:00 a.m. Dallas-to-New York flight has an average of 10 percent no-shows at any given time. (And now those of you who have railed against overbooked flights know why—algorithms are good, but not perfect.)

But after that it got more sophisticated. What if they noticed that people book and then cancel within twenty-four hours, or people cancel within whatever time cutoff? The airlines created cancellation forecasting so they can take even more bookings than the capacity before the flights take off because they know so many of them are going to cancel.

Then the airlines started thinking, well, people buy at different price points. They started segmenting consumers into those different price points. They call them buckets. They started forecasting the *demand* for every bucket and how many people are typically in this bucket. So, the absolute bargain fares made last-minute to places like Las Vegas are very different from first-class flights from LA to Hong Kong.

Now, I'm going to pause here for a moment. It can sound like all you do is plug some numbers in a machine and *voila,* the airlines can perfectly sell and book their planes. But it's not that simple. (Hence, overbooked flights!)

For example, looking at those buckets we just discussed, the airline will try to accurately know how many people there are willing to pay high prices for premium seats. If you have more people that are able and happy to pay a higher price to be more comfortable, you want to save seats for them rather than cut prices to just fill them. The algorithms also predict that those willing to pay higher prices may come at different points in time. For example, business travelers flying off to deal with a crisis at a headquarters overseas may book last minute whereas a large family planning Christmas in Switzerland may book as much as nine months ahead.

Depending on your age, you may remember the discount People Express Airlines offered. It went out of business, partially because it failed to adapt to an evolving competitive landscape, which increasingly included using RM systems. The insights that I was describing, forecasting by bucket, deciding which inventory, enabled its competitors, such as American Airlines, to overtake the market. So, although People Express had extremely low-cost fares (it was arguably the earliest low-cost carrier post-deregulation), it lost to high-cost airlines.

Wait. Why would a low-cost airline lose to a high-cost airline that has unions, high-salaried pilots, etc.? Because although People Express was a low-cost airline, it was filling up on the cheap passengers—which meant that they couldn't even cover their own costs. As an example, American Airlines, with much higher costs, didn't need to slash fares across the board because it had (and has) all these forecasting algorithms and optimizations, which allowed it to offer inventory at the right time, for the right price, for the right consumer. This RM discipline of decisioning is an analytics framework driven by algorithms that conduct demand forecasting, no-show forecasting, cancellation forecasting, inventory optimization, and more.

This airline case study has been used in marketing classes and business schools across the world. It shows what data is capable of and why, perhaps, it's even capable of sinking an airline.

Fast forward to the present, and the airline industry has embraced becoming a data-driven business, mirroring the innovative paths of other industries. Central to this transformation is the frequent flier program that airlines maintain. It is a story of every frequent traveler, whose every flight, seat preference, destination choice, and in-flight purchase are meticulously recorded and analyzed. Furthermore, this database is further enriched by third-party data to have a more comprehensive view of every frequent flier. Then, leveraging data analytics platforms, sophisticated models are built to understand and predict customer behavior with precision.

The implications of such data operations are profound. For airlines, where the competition is fierce and the margins often tight, ancillary revenues generated through targeted marketing and strategic partnerships have become a critical success factor. The data gleaned from millions of travelers enables airlines to not only tailor their service offerings but also create a symbiotic relationship with credit card companies, luxury resorts, and other partners, offering exclusive deals that benefit all parties involved. Furthermore, this approach allows airlines to refine their loyalty programs, offering tiered rewards that genuinely reflect the preferences and behaviors of their most valued passengers. It's a dynamic ecosystem where data is continuously enriched and reevaluated to ensure the highest levels of customer engagement.

Jonathan's travel patterns, from his seasonal trip preferences to his loyalty program activities, are aggregated to form a rich profile picture.

This isn't just about ensuring Jonathan's next flight meets his expectations; it's a cornerstone of a much larger strategy. By integrating frequent flier data with partnerships from airline credit cards to hotel and rental services, airlines can offer Jonathan personalized deals that are hard to resist. These offers are not random but are the result of analyzing vast amounts of data to understand what Jonathan values most.

RETAIL INDUSTRY EXAMPLE

Loyalty cards are another data creep. What started as a way to give out coupons has become much more intrusive. By knowing what you buy every time you shop, companies can extrapolate much more information than the shopper is aware of.

You recall Erica, whom we introduced earlier? She is a mom managing the bustling life of a family of four but represents more than just a consumer in the eyes of today's retailers (and other companies). Her morning ritual of leveraging loyalty cards for purchases, like those Eggos for her toddler, unfolds a narrative deeply intertwined with the evolution of consumer data collection and utilization in the retail sector.

In the 1980s, a groundbreaking shift occurred with Catalina Marketing pioneering the integration of loyalty programs and targeted coupons. (Full disclosure: I worked for Catalina at one point.) This initiative marked the beginning of a data-driven transformation in retail, mirroring the sophistication seen in airline RM systems. Catalina's innovation at the time laid the foundation for retailers to not just sell products but to understand consumer behaviors, preferences, and patterns at an unprecedented level.[27]

Fast-forward to the present, and companies like Kroger have taken this data-driven approach to new heights. Kroger's subsidiary 84.51° stands as a testament to the retailer's commitment to leveraging

---

27   https://www.company-histories.com/Catalina-Marketing-Corporation-Company-History.html.

data science and analytics. Named after the longitudinal location of its headquarters, 84.51° is composed of a legion of analytics and data science professionals dedicated to creating detailed consumer profiles. These profiles are not merely for enhancing customer experience but serve as a vital component of Kroger's retail media network business and other data-centric ventures.[28]

The significance of such data operations cannot be overstated. For retailers, using our example Kroger, and other grocery chains, the margins from selling groceries are often slim, reflecting the challenging nature of the retail business. However, the data business offers a lucrative avenue. By analyzing and monetizing consumer data, retailers can unlock new revenue streams that are often more profitable than traditional retailing. The insights derived from consumer behaviors—what they buy, when they buy it, and the offers that influence their decisions—enable retailers to not only optimize their own marketing strategies but also sell these insights to brands looking to target consumers effectively and much more.

It's crucial to recognize that Kroger is not alone in this venture. Other retailers, from large chains to niche markets, have recognized the value of data and have embarked on similar paths. They've established or are in the process of developing their own data analytics capabilities, contributing to an industry-wide ecosystem where consumer data is as valuable—if not more so—than the products on the shelves.

Erica's story, from her early morning tea routine to her strategic use of loyalty cards and coupons, encapsulates a broader narrative of how retail has evolved. It's a world where every transaction tells a story,

---

28   "Kroger's 84.51° launches new tool to capture customer behavior," 2019, https://progressivegrocer.com/krogers-8451deg-launches-new-tool-capture-customer-behavior#:~:text=84.51%C2%B0%20has%20been%20around%20since%20 2015%2C%20when%20Kroger,named%20for%20the%20longitudinal%20 location%20of%20its%20headquarters.

every purchase feeds into a vast data network, and every consumer action shapes the future of retail strategies. Just as the airline industry was transformed through data analytics, the retail industry's journey into data sophistication illustrates a parallel evolution, highlighting the transformative power of data analytics.

BEYOND AIRLINES AND RETAILERS: ALL INDUSTRIES

The widespread utilization of data is across various industries and goes far beyond just airlines and retailers. It is a universal adaptation of data analytics in the digital age. While the examples of the airline and retail industries provide a window into the transformative power of data analytics and consumer profiling, they represent just the tip of the iceberg in the digital age's data story. Virtually every savvy business across any sector/industry leverages data to tailor services and products to their consumers' needs, evidencing a universal shift toward data-driven decision-making.

In the automotive industry, data analytics guides everything from vehicle design based on consumer preferences to personalized marketing strategies for potential buyers. Insurance companies utilize vast datasets to adjust policy offerings in real time, pricing premiums more accurately by assessing individual risk factors. Banking and financial services have transformed customer engagement through personalized financial advice, risk assessment, and tailored product offerings based on transaction histories and online behavior. The entertainment industry uses viewing habits to recommend content, creating highly personalized viewing experiences. And many more!

This wide-reaching embrace of data underscores a fundamental truth: in the digital age, savvy businesses across all sectors recognize the critical role of data in achieving competitive advantage. It's a testament to the era where understanding every individual intimately through data analytics is an advantage and a necessity for both survival and growth. However, this relentless pursuit of data-driven insight comes at a significant cost to individual privacy. As consumers, we often remain blissfully unaware of the extent to which our data is stitched together from various sources, painting a comprehensive picture of our lives, preferences, and behaviors for those willing to pay for it.

Every click, purchase, location check-in, social interaction, *every* interaction, and *every* transaction contributes to an ever-expanding digital dossier that businesses mine and monetize. This constant monitoring and analysis of our digital footprints have led to a profound erosion of privacy, with personal details no longer confined to the realm of the personal. Our data is a commodity of immense value, fueling a digital economy where individual privacy is often the price of admission. The worth of this data is not just in the billions but trillions, driving industries and economies forward at the expense of personal autonomy and privacy. Hence, at the expense of freedom!

As we navigate this digital landscape, the challenge becomes striking a balance between leveraging technology for convenience and innovation, while safeguarding our right to privacy. It raises critical questions about data ownership and independence, data consent, data control, data currency, and the ethical use of personal information. The digital economy thrives on data, but it's imperative that this does not come at the cost of individual privacy rights being overlooked or undervalued. In this new era, understanding and protecting the value of our personal data is just as crucial as the businesses that seek to profit from it.

REVISITING OUR PRIVACY TIMELINE BY FOLLOWING THE MONEY

At first glance, we can look at the evolution of our loss of data privacy with the gains of Big Data. Let's delve into each stage again and provide some numerical data regarding company valuations and growth. It's important to note that the specific numbers can vary depending on the source and the date of the valuation. The figures provided will be representative of the trends during the respective periods. This is by no means exhaustive. My intent in providing this is to show readers just how much value your data has to others—yet, last I checked, you were not reaping any monetary benefits, nor was I. (Alas, I am not retiring to the south of France.) Nor do we have much choice at the moment as to who gets to reap those benefits.

Centralized Databases and Mainframes (1960s through 1970s)

In the late 1960s and 1970s, IBM dominated the computer industry. By the early 1980s, IBM's market value had reached nearly $70 billion, making it one of the most valuable companies in the world at the time.

Credit Reporting and Databases (1970s through the 1980s and Beyond)

Equifax, one of the largest credit bureaus, saw its revenue grow significantly during this period. By the 1980s, the company was already making hundreds of millions in revenue. Its revenue in 1993 was $5.1 billion.[29]

Loyalty Programs (1980s and Beyond)

The exact valuation increase due to the AAdvantage program is hard to quantify, but American Airlines' market cap in the late 1980s was several billion dollars, with loyalty programs contributing significantly

---

29   "Equifax," https://www.forbes.com/companies/equifax/?sh=3d0673e539e8.

to customer retention and revenue. As I wrote in chapter 1, we increasingly see loyalty programs with higher valuation than companies' physical assets.

Personal Computers and the Internet (1980s and 1990s)

By the end of 1999, Microsoft's market capitalization had soared to a staggering $600 billion, making it one of the most valuable companies in the world at the time. Founded in 1994, Amazon's market cap had reached about $500 million by 1997, and it continued to grow exponentially in the following years.

Online Advertising and the Era of Cookies (1990s through 2000s)

Google's IPO in 2004 valued the company at $23 billion, and by 2007, its market cap had grown to over $200 billion. Facebook was founded in 2004, and by its 2012 IPO, Facebook was valued at $104 billion.

Social Media (2000s through Today)

Facebook is one of the "Big Five" digital overlords in the United States—Meta, Twitter, Apple, Google, and Amazon. Its market cap surpassed $1 trillion in January 2024.[30] Twitter, launched in 2006, had a market valuation of about $31 billion at the time of its IPO in 2013.

---

30   Alex Koller, "Meta passes $1 trillion in market cap," 2024, https://www.cnbc.com/2024/01/24/meta-passes-1-trillion-in-market-cap.html.

Smartphone Location Data (2000s on)

Apple's market cap crossed $300 billion in 2010, driven in part by the success of the iPhone (and all that data!).

Internet of Things (IoT) (2010s on)

By 2018, Amazon's market cap had briefly crossed $1 trillion, boosted in part by its diverse product range including IoT devices like Alexa.

Big Data, Machine Learning, and Artificial Intelligence (2010s on)

In 2016, NVIDIA's market cap was around $14 billion, but by 2021, it had surged to over $300 billion, reflecting its leadership in AI.

These numbers are indicative of the monumental growth in valuation experienced by key players in each stage. They reflect how advancements in technology and data management directly contributed to the financial success of these companies.

SOME OF THE MISCONCEPTIONS WE HAVE BEEN TOLD

As this data creep has occurred, the public and individuals, media and government, have been offered some outright (at best) fallacies and misconceptions by the data collectors. Here's a small handful (and the reality).

If you have nothing to hide, you have nothing to fear. This is one of the more absurd arguments I have seen (and certainly would not be posited by anyone concerned about privacy). I may have nothing to hide in my house either, but if you come to it wanting to take a look around, you had better have a warrant.

Data collection is only for targeted ads. While this may be *one* use for data, it is not the only one. As we have learned, our data can be used in ways most individuals have no idea about. Targeted ads seems like an innocent time compared to where we are now!

Anonymized data protects your privacy. Even if it is, as we explored in the chapter, a clean room—combining one dataset with another and correlating it—removes that anonymity. *They know who you are.*

This app needs to access your location/photos/contacts in order for it to function. Sometimes this may be true—but often more data is collected than needed. And there will always be bad actors.

I know some people believe "incognito" mode and virtual private networks will protect their data—not so. Deleting apps doesn't protect you either.

Welcome to the Wild West of data collection.

CHAPTER RECAP

Who is reaping the rewards of your digital breadcrumbs? In this chapter, we examined the data "creep"—and the nascent origins of algorithms in predictive consumer modeling.

And they are deeply unsettling.

However, I am excited to bring you to this next chapter.

I believe there is a solution. It is not going to be found in legislation (and I will explain why in greater detail later, but suffice it to say loopholes are exploited)—but it *will be found* in a new way for us to look at our data, the ownership of that data, and how we (yes, us as individuals) can use it.

CHAPTER 4

Breaking Free: Is Individual Sovereignty in the Digital Landscape Possible?

Arguing that you don't care about privacy because you have nothing to hide is no different than saying you don't care about free speech because you have nothing to say.

—EDWARD SNOWDEN

I could never do what I do without finding solutions.

In the grand narrative of science and exploration, each individual's contribution is akin to a single thread in a vast and intricate tapestry. My research scientist role at the Center for Space Research is a very humble portion of this grand design, crafted by the collective intellect and creativity of countless brilliant minds. In the endeavor of landing a rover on the uncharted terrains of Mars or guiding our courageous astronauts back to Earth, it becomes clear that these milestones are not the triumphs of a lone genius, but the collective achievements of a united dream, a shared pursuit of the unknown. Every mistake, every detour in our scientific journey, contributes essential shades and contours to the canvas of knowledge. We find ourselves walking a path illuminated by the pursuit of truth, where each so-called failure is a beacon, revealing new directions and possibilities. Our stumbles and missteps are akin to the thoughtful reflections of Thomas Edison, whose vision saw these moments not as setbacks but as invaluable lessons guiding us toward future success.

Harnessing the same spirit of collective ingenuity and shared endeavor that propels our scientific milestones, let us apply this collaborative ethos to tackle the multifaceted data challenges we face today. Despite our advances and achievements, we are acutely aware of the complexities and challenges that persist in our relationship with data. Let us recognize that there's a data problem encompassing privacy concerns, where individuals' information is often vulnerable and exposed.

We also have the issue of data control, highlighting the need for people to have greater autonomy over their own information. Furthermore, the monetization of data presents ethical and practical dilemmas, questioning the fairness of profiting from personal data without equitable benefit to its owners.

In light of these ongoing and evolving challenges, I propose not just a remedy but a starting point for dialogue. This suggestion is intended to build upon our collective wisdom garnered from years of navigating the data landscape. It's a call to unite, to leverage our shared experiences and insights, to forge a path forward that addresses these critical issues. Our goal should not only be to innovate but to do so responsibly, ensuring that our advancements in data utilization are matched with strides in privacy, control, and equality. Let this proposal serve as an invitation for all stakeholders—researchers, policymakers, industry leaders, and the global community—to collaboratively develop solutions that uphold the integrity and dignity of every individual in the digital age.

This journey of exploration and understanding is not just confined to the realms of space and science. It extends to the intricate world of data, a domain where I am venturing into new territories.

But before we talk about my solution (which will be fully revealed in part II), I want to define a few terms and topics you may hear frequently in discussions of data. These few terms and topics are not meant to be a comprehensive technology review of data but essential components to discuss and convey some important points.

The Clean Room. As we touched upon in the previous chapter, the clean room is shrouded in a veil of illusion. It's portrayed as a "top-secret room" where companies supposedly guard your privacy while crafting a 360-degree view of you. Imagine various entities that are observing every Google search, every purchase, activity, etc., and deciphering your identity through an ID Graph. All this information needs to be put together so as to have a continuous 360-degree picture of you, the individual. But here's the twist: these rooms are not as pristine as they seem. In reality, they are a clever masquerade, a sophisticated dance of data manipulation so as to achieve this goal.

Here are some typical steps of how it unfolds: Companies possess PII, which they obscure using a hashing algorithm, creating a façade of anonymity. This data, masked yet rich in detail, is then transported into the clean room. Inside this digital chamber, companies match their hashed data with that of other entities using the same hashing technique. This clandestine ballet of information allows them to intertwine and compare datasets, revealing insights without explicitly sharing raw PII.

However, this is not a sanctuary of privacy. It's a meticulously crafted loophole, a testament to how the art of lobbying forges pathways to protect the lucrative gains of the Data Economy (our new currency). These clean rooms, while legal, are a subtle insult to the intelligence of the public, a veneer of privacy over a robust engine of data exploitation. This is but one example of how the architecture of our digital world is engineered to safeguard commercial interests, ensuring the flow of capital in the data-driven market remains uninterrupted and unchallenged.

Walled Gardens. While gardens may conjure up positive connotations, in the data world, these are how our digital overlords maintain control (see figure 4.1). They are formidable technology fortresses, segregating vast territories of digital landscapes. In these realms, technology giants like Google, Apple, Amazon, and social media behemoths like Meta, major retailers, and many more companies reign supreme. These spaces are meticulously walled off, preventing interaction with applications or content from external sources.

As an example, in the digital advertising world, the concept of walled gardens takes on a more profound meaning. These big players, because of their massive scale, have accumulated extensive data on their users. They have no incentive to share these data troves with competitors, but they might still enrich it with some assets they don't have or collect. Essentially, they already possess a wealth of information, rendering any external data superfluous. As a result, users are confined within these digital walls, often unknowingly.

But the implications of these walled gardens go beyond just data isolation. They represent a new era of data monetization. Retail media networks are a prime example of this.[31],[32] These networks allow retailers to leverage their customer data to offer targeted advertising opportunities within their own properties and platforms. For instance, a retailer like Walmart or Amazon can use the purchasing history and browsing behavior of their customers to provide targeted

---

[31]   Sara Lebow, "Retail media networks: definition, benefits, types, and examples," emarketer.com, https://www.emarketer.com/insights/definition-retail-media-networks/.

[32]   Jason Goldberg, "Retail media networks are one of the most important trends of 2022, but they need to evolve," *Forbes*, 2022, https://www.forbes.com/sites/jasongoldberg/2022/02/28/retail-media-networks-are-one-of-the-most-important-trends-of-2022-but-they-need-to-evolve/?sh=67b113fc4516.

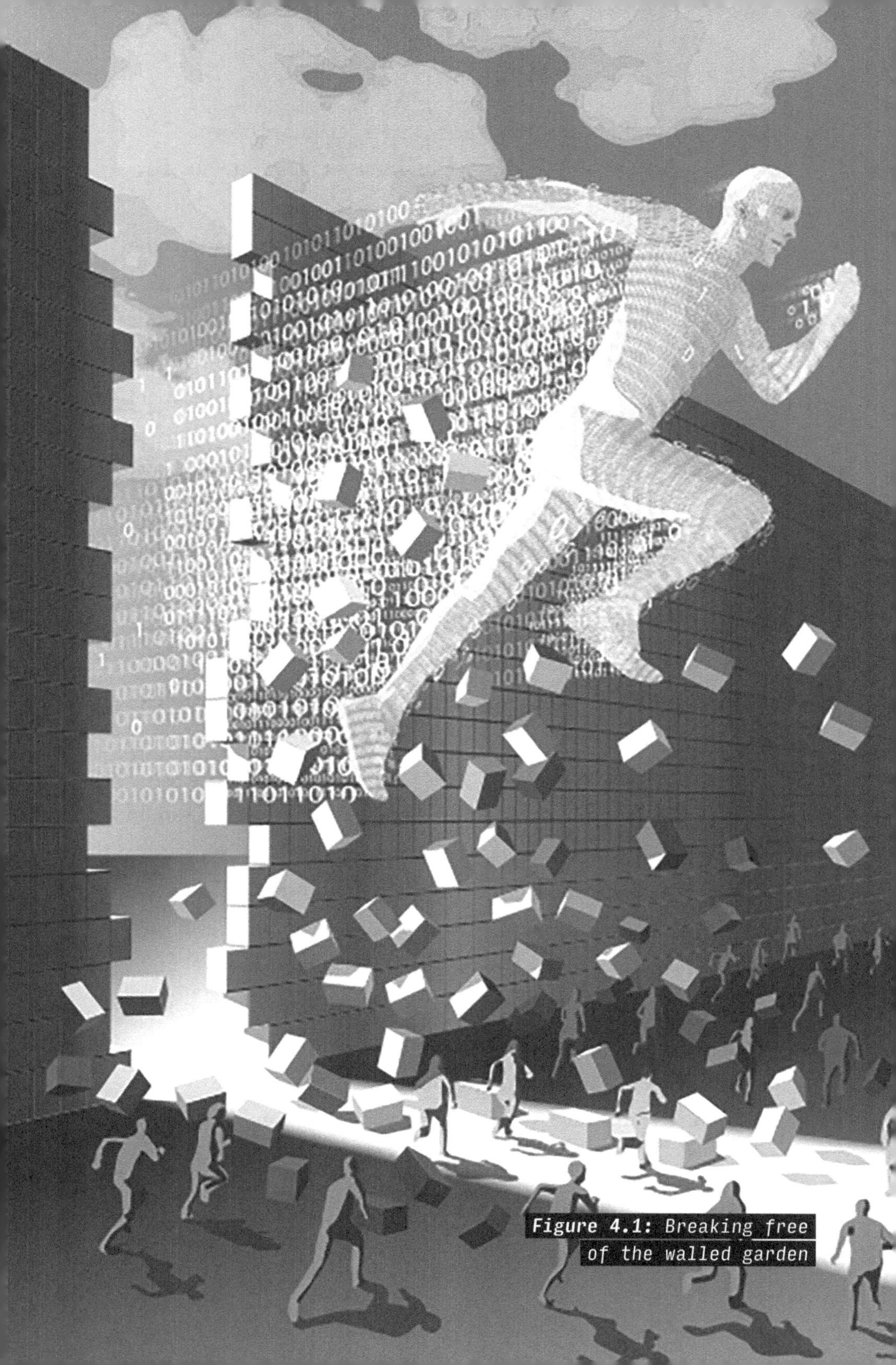

Figure 4.1: Breaking free of the walled garden

ad placements for brands within their properties and platforms. This not only creates a new revenue stream for these retailers but also offers advertisers a more precise way to reach potential customers.

Furthermore, walled gardens are not just limited to retail. Streaming services like Netflix or Spotify also operate within their own walled spaces, using subscriber data to personalize content and advertisements. Social media platforms leverage user interactions and preferences to sell targeted ad space. The scale of data these companies possess makes them invaluable players in the digital advertising landscape.

In all these cases, the common thread is the control and monetization of data. The individual user, often unaware of these capabilities, remains outside the decision-making process. Data privacy becomes a casualty in these scenarios, as most people are oblivious to the extent of data collection and utilization happening within these walled gardens.

This new digital era, dominated by few but impacting many, raises critical questions about data ownership, privacy, and the ethical use of personal information. It underscores the need for greater transparency and user consent, control in the digital world, where personal data has become a currency more valuable than gold. Any individual should have the key to their own data's consent, control, and currency (C3).

Cookies. One concept that gives users an illusion of control is cookies. These are small text files that are placed on an individual's computer when they visit a website, which then customizes their website experience. The next time the user visits that site, their preferences will be stored so that their experience is more seamless.

For example, if I like to shop at J. Crew and leave two sweaters in my shopping bag, the next time I visit the site, it may remind me I still have those two sweaters sitting there, just waiting for my credit card.

The term "cookies" is also not monolithic. There are temporary cookies (called "session" cookies). Next are first-party cookies (from the website itself, so for example a session cookie from J. Crew would be from J. Crew's website itself). There are also third-party cookies ... http-only cookies. You get the idea.

Another simple example is when you return to a website you have shopped at before and your name and information auto-populate the mailing or billing form. Cookies can set your preferred language, authenticate your identity, or track your ads.

We have become so used to the ubiquitous pop-up windows asking if it is all right to use cookies—and whether we opt in or opt out, many innocent consumers and individuals are unaware just how intrusive ad-tracking cookies can be—those that follow us merrily from site to site. They follow users—not to make our experiences online better and more seamless but to make companies more money. It comes down to greed, pure and simple.

They are stitching us together—and this happens invisibly to us, largely without our understanding and usually without our consent.

Legislation. Another illusion of control is legislation. We will discuss this in detail in part II, but when have big pieces of legislation about *anything* solved huge problems? The fact is, the data economy, Big Tech, and other companies that are data-driven businesses are worth trillions, all having very, very brilliant, very, very expensive lawyers on retainer, searching for loopholes to exploit.

For example, GDPR in Europe may have had good intentions— but its bloated enforcement processes mean that only 65 percent of, for example, Ireland's complaints under GDPR had been handled at the time of a 2022 *Wired* article.[33] Other enforcement actions have dragged on for years.[34]

---

33   Matt Burgess, "How GDPR is failing," *Wired*, May 23, 2022, accessed December 28, 2023, https://www.wired.com/story/gdpr-2022/.

34   Ibid.

Some may remember one of Mark Zuckerberg's Senate appearances, where he had to give Senator Orrin Hatch a response to the world's biggest softball question: How does Meta make its money. "We sell ads," was Zuckerberg's response. Which was accepted at face value by Hatch.[35] Not one follow-up adequately addressed the algorithms—that *really* are how and why Meta makes its money.

In 2017, while studying law, Lina Khan gained attention with her law review article "Amazon's Antitrust Paradox," critiquing the outdated approach of US antitrust enforcement. She argued for a shift beyond just consumer welfare and price-focused actions, especially relevant in the era of digital giants like Amazon and Facebook.[36]

Four years later, in 2021, President Biden appointed Khan as chair of the Federal Trade Commission (FTC), a key agency in US antitrust enforcement. This role enabled her to start implementing her forward-thinking ideas.

Despite these efforts, the FTC has experienced setbacks in court. I admire the FTC's efforts in tackling these complex issues. However, the challenges we face today demand a transformative approach that redefines the ownership of data, placing it firmly in the hands of individuals. This paradigm shift should emphasize consent, control, and currency of personal data, enabling each person to have a "data vote." Such a move would not only empower individuals in our democratic society but also guide the future trajectory of technology and AI advancements, ensuring they align with the collective will and benefit of the people.

---

35 Mathew Ingram, "Congress fails to grapple with social networking algorithms," *Columbia Journalism Review*, April 29, 2021, accessed December 17, 2023, https://www.cjr.org/the_media_today/congress-fails-to-grapple-with-social-networking-algorithms.php.

36 Lina M. Khan, "Amazon's Antitrust Paradox," *The Yale Law Journal* 126, no. 3 (2016–2017): 564–907, https://www.yalelawjournal.org/note/amazons-antitrust-paradox.

Empowering individuals with their data vote is not just about personal empowerment; it's a strategic move toward resolving deep-rooted anticompetitive issues in the digital marketplace. By granting each person control over their data, we effectively decentralize the power that currently resides in the hands of a few dominant tech and business giants—the digital feudal system. When individuals have the authority to decide who can access their data, through a seamless paradigm, and for what purpose, it levels the playing field. Companies will no longer be able to leverage vast reservoirs of data unilaterally but will instead need to earn the trust and consent of each user. Companies will have to compete for every individual's "data vote."

A fundamental shift in data dynamics is needed! This shift in data dynamics fosters a more competitive and diverse market landscape, where companies succeed not by hoarding data but by offering genuine value and respecting user preferences. In essence, the data vote becomes a powerful tool in the hands of the people, enabling them to actively shape the market, shun away the anti-competitive players, and champion fair, ethical practices in the use of technology and AI.

So, if privacy is freedom, is our only choice to roll over and give our freedoms away?

No.

There is a solution.

THE PERSONAL DATA VAULT KEY (PDVK) SOLUTION

Remember Apple's groundbreaking 1984 Super Bowl ad that depicted the runner hurling a sledgehammer in a dystopic world and promising us a new way of computing? Figure 4.2 depicts a new breaking free.

Figure 4.2: Freedom achieved through the PDVK

I am suggesting a new way of *owning back* our data, a new way of creating ownership not only of our own data but insisting others take ownership of their data so there will be no "fake news" or fraud because every piece of information will come with an ownership ID or key. We will learn more about this paradigm shift shortly—but first, I want to explore the very foundational principles upon which we would rest a solution to our data privacy crisis.

CHAPTER RECAP

In this chapter, we discussed concepts like cookies—which give an *illusion* of control but actually do nothing to protect and honor your data privacy.

We also discussed clean rooms and walled gardens—and their impact on our online lives (and the mercenary nature of them). We also explored how legislation largely leaves us defenseless against the digital overlords. We are mere serfs.

We are going to go into the details of what I am calling a PDVK in the next two chapters—launching us into part II. The principles of PDVK are rooted in the consent, control, and currency (C3) foundational trio. This represents the core rights of individuals over their data—to give informed consent, exercise control, and benefit from their data as a form of currency.

In part II, we will discuss not only C3 but also the five main principles of the PDVK, which are described through the acronym TAPER.

It's time to declare our data independence!

PART II
Our Data
Independence

CHAPTER 5

The Principles of the PDVK: The C3s and TAPER

Ethics is knowing the difference between what you have a right to do and what is right to do.

—POTTER STEWART

Now that I have briefly introduced the concept of our PDVK, which we will go over in greater detail in chapter 6, let us first discuss the principles on which it is based. These are principles, as I spell out, that can be a stronger foundation for how we fundamentally view our data and digital lives.

For most of our lives, we have thought of our data as something rather passive. We put terms in a search engine; we book a plane ticket online; we order books, cosmetics, and clothes from Amazon; we shop for groceries at the store; Uber Eats or DoorDash; we use our GPS to find the new restaurant we chose for dinner; we book rental cars and Airbnbs. Whatever it is we are doing offline or online, despite whatever basic security measures we put in place (such as declining cookies or being cautious about who we give our email to), is a digital footprint or breadcrumb that can be exploited.

When I considered what I wanted the PDVK to accomplish, I realized that, just as our forefathers considered life, liberty, and the pursuit of happiness as they crafted their vision for the future of our young nation, we would have to consider just how our data impacted our privacy and freedom now and in the future.

Thus, even as I was devising the concept of the PDVK, I had to pause. The PDVK could not be a series of governmental or corporate legal-ese tying itself into knots with complicated verbiage, but *principles* we all could understand, *principles* that protect our PDVK. Not people in the tech world, not Silicon Valley—but the average citizen who is just now beginning to understand the power of their data.

In part I of the book, we touched on our C3s: to give informed *consent*, to exercise *control*, and to benefit from our data as a form of *currency*.

Part I made clear that not only is our data being collected far more than most of us think it is, but we are also being manipulated, studied, and stitched together in the digital world without our complete understanding. We thus need a paradigm shift when it comes to our data. It is ours, and it is ours to consent to its use and how it is used. We should be in control of our data—it is *ours*. In addition, our data is worth not hundreds of millions of dollars, but trillions, collectively. Yet, it is the data collectors who benefit from our digital DNA. They monetize our digital DNA. Why are we not reaping benefits from our own data? It is a form of currency. Forget crypto. Data is king. (Or queen!)

The diagram in figure 5.1 will give you an idea of how the C3s work in conjunction with our new acronym: TAPER. This stands for transparency, accountability, protection, equality, and respect. We will discuss these five areas in the rest of the chapter.

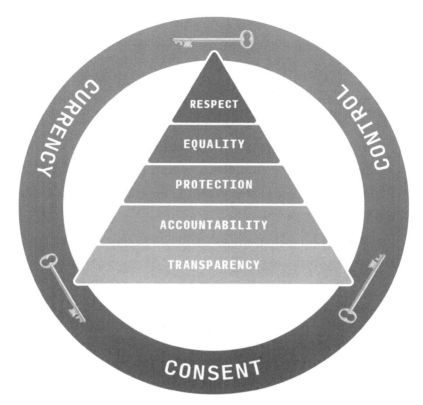

Figure 5.1: The C3s and TAPER

THE FOUNDATIONS OF TAPER

The national motto of France is "Liberty, Equality, Fraternity." Those concepts are the foundations of its values, just as America loves her freedom. We must have principles we can all agree on for our data independence in the future.

As the illustration makes clear, TAPER and the C3s work together. Whatever solution the world comes up with when it comes to data privacy, whether it is the PDVK or some other novel solution, it will need to address the foundational pillars of the paradigm shift.

I created the graphic using a pyramid—because only by building on these foundations will we have strength both individually and collectively in the new data economy.

> **Transparency (T)**: Ensuring clear and open practices in data usage, storage, and sharing.
> **Accountability (A)**: Entities handling data must be accountable for their actions and compliance with these principles.
> **Protection (P)**: Strong measures to safeguard data against unauthorized access and breaches.
> **Equality (E)**: Guaranteeing that all individuals have equal rights and opportunities in the digital realm, irrespective of their background or abilities.
> **Respect (R)**: Upholding the dignity, privacy, and rights of individuals in all data-related processes.

Let's address each of these in turn. First up is transparency.

Transparency

You may not recall the epigraph that started this book, but Steve Jobs is quoted as saying, "I believe people are smart. Some people want to share more than other people do. Ask them."

Jobs was a visionary, but I don't believe that most people know what they are consenting to when they are asked in the digital world. But they should, indeed, be asked. We need more basic data privacy literacy. We often hear about "opaque" data collection. This is about transparency. Are companies telling their clients, users, and the like the extent of the data they are collecting?

What about the algorithms that seem to govern so much of our online lives? Lack of transparency there can lead to unconscious biases and an undermining of trust.

In a survey of 1,580 executives from large organizations across ten countries, and over 4,400 consumers across six countries, "Among consumers surveyed, 62 percent said they would place higher trust in a company whose AI interactions they perceived as ethical. ... By contrast, when consumers' AI interactions result in ethical issues, it threatens both reputation and the bottom line: 41 percent said they would complain in case an AI interaction resulted in ethical issues, 36 percent would demand an explanation and 34 percent would stop interacting with the company."[37]

Another area that lacks transparency is data monetization. Most consumers are aware our data is collected and is worth "something" (with little awareness of how valuable it really is). However, we tend to believe it when companies tell us they will not "sell" our data to others. However, what *are* they doing with our data, and how can it be monetized?

When a retailer of any sort collects our data, there should be transparency of why they need this data to conduct commerce. Remember, I am not trying to cripple commerce. If I go in-store to buy or online and order a sweater, I necessarily have to provide information on how to deliver to me, as well as my credit card or payment information.

Transparency as I mean it here, in terms of commerce, is that if a retailer needs our data to conduct commerce that we have chosen, that is acceptable. (I have *consented* to that.) If they "need" it because they want to have that 360-degree stitched-together view, they should not have the right. (I did not consent to that, and the data is no longer in my *control.* In addition, if my data is sold, I have lost out on that *currency.*)

Just because Jonathan bought a plane ticket and Erica bought dog food online does not mean that retailer needs their Social Security number. No, the purchaser does not provide their actual Social Security number (at least I hope they do not!), but the retailer still

---

37 "How consumers view the transparency of their AI-enabled interactions," 2019, https://www.helpnetsecurity.com/2019/07/11/ai-enabled-interactions/.

has their ID Graph with a fully developed view of everything about our characters (and about you and me too).

If I go into Macy's and offer them my credit card to purchase new sunglasses, the transaction should be completed and that should be the end of it. But that never is, is it? Retailers from the grocery store to Big Box stores to clothing retailers, etc., will ask you for your email and your phone number. This is to "stitch" together our profiles to have a 360-degree view of us, the better to manipulate and sell us things, the better to analyze our behaviors and sentiments and use them against us.

Meanwhile, I am offering examples from grocery stores and retail. However, as we learned in part I, there is the selling of intimate data on a larger scale, to any entity, that is occurring. Transparency can have even deeper importance when it comes to personally treasured freedoms, such as where we worship, what political party we donate to, or what books we buy.

A *Politico* feature on journalist Bryan Tau's discoveries while researching his book, *Means of Control,* referred to: "An opaque network of government contractors is peddling troves of data, a legal but shadowy use of American citizens' information that troubles even some of the officials involved. And attempts by Congress to pass privacy protections fit for the digital era have largely stalled." (We will later discuss how legislation has largely failed on this issue.) "Any nightmare use for data you can think of will probably eventually happen," Tau said. "It might not happen immediately, but it'll happen eventually."[38]

With the PDVK, transparency will be a given—no more second guessing of where our data is and how it's being used—because we will be in control.

---

38 Steven Overly, "The government really is spying on you—and it's legal," *Politico*, 2024, https://www.politico.com/news/magazine/2024/02/28/government-buying-your-data-00143742.

Accountability

Next in our TAPER acronym is accountability. Entities handling data must be accountable for their actions and compliance with these principles. Under the PDVK, there will be data brokers—but there will be far more accountability than we demand of those buying and selling our data today.

An example I use is if I give Fidelity Investments my money, they are responsible for investing it the way I have instructed them. They're accountable. They're responsible for it. While I cannot control the markets, I *can* control how I invest and make choices that align with both my financial goals—and my personal beliefs.

Consider this: If my money manager at Fidelity invests in Big Box Store, and that Big Box Store does something very unethical, and there is a lawsuit, Big Box is accountable to us all. If I am not impressed with Big Box's response to this ethical lapse or issue, I can tell my money manager to pull my investment from Big Box. I am "voting," if you will, with my dollars.

Another example of consumer "voting" is the boycott. Some consumer boycotts have worked and pressured corporate entities to change their policies. Harry Winston, after a public pressure campaign, announced they would no longer utilize diamonds from Burma, in consideration of its human rights violations.[39]

One of the most successful boycotts in history occurred when Cesar Chavez demanded change for farm workers and millions boycotted grapes in 1965.[40] Over and over throughout the years,

---

39   "History of successful boycotts," 2023, https://www.ethicalconsumer.org/ethicalcampaigns/boycotts/history-successful-boycotts.

40   Michael Livingston, "Here's When boycotts have worked—and when they haven't," *LA Times*, March 1, 2018, accessed February 14, 2024, https://www.latimes.com/nation/la-na-boycotts-history-20180228-htmlstory.html.

consumers have voted with their wallets, sometimes changing the world (sometimes not).

But why shouldn't we be able to vote with our data as well? Why shouldn't we be able to ask those who use our data to be accountable to us? Our data is just as much an asset as our money.

It is the *new* currency.

Protection

Next, we come to P for protection.

You may recall from part I where we explored some of the biggest data breaches in history from Equifax to Yahoo. We hear about data breaches almost daily. Even more chilling, as our lives evolve to be increasingly online, everything from pictures of the inside of our homes (on Zillow and the like), to our DNA (23andMe), to our entire health records, our employment files, all of our financial records, etc., is stored by many entities.

And bad actors know that.

After Edward Snowdon revealed the existence of the very secretive Bluffdale NSA data center in Utah, the NSA data center there began experiencing 300 *million* attacks each *day.*[41] Hackers and bad actors abound. If 166,000 white hat hackers are registered in the ethical HackerOne—we can only imagine how many unethical hackers there are across the globe.[42] Hackers are responsible for 45 percent of data breaches that occur.[43]

---

41 "NSA Data Center effect: This state experiences 300,000,000 hacking attacks a day," 2023, https://www.techworm.net/2016/02/nsa-data-center-state-hacking-attacks.html.

42 Jason Wise, "How many hackers are in the world in 2024?" 2023, https://earthweb.com/how-many-hackers-are-in-the-world/.

43 Trevor Cooke, "How many data breaches in 2022? (2024 statistics list)," 2023, https://earthweb.com/how-many-data-breaches/.

It goes without saying that any solutions to our data problems must include strong measures to safeguard data against unauthorized access and breaches. In addition, if the government can buy and access commercial data, what is to stop foreign adversaries from accessing our data today, or the data of its citizens.

Thus, just as Fidelity or other financial institutions must have extraordinary safeguards in place, so too must data brokers and anyone who collects our data for any purpose. Fortunately, when we begin talking about the PDVK in the next chapter, it will become clear that the key will provide protections that will help identify these bad actors.

We must add to or strengthen existing laws to take into account that we are in the digital economy and the digital world. Just as we safeguard our bank accounts and our DNA, so too must our digital DNA be protected. In addition, crimes committed against us in terms of hacking and data breaches must be seen for what they are. Attacks on our digital DNA—the secrets within our own digital double helix—must be prosecuted as such.

The protections I am suggesting in the PDVK are important protections for all of us, in terms of new assaults on our digital lives, such as deep fakes. The 2024 horrifying release of AI-generated explicit pictures of Taylor Swift should send a chilling effect through each and every one of us. At a New Jersey high school, fake nudes of fellow students in group chats sent a shockwave through the school—and parents.[44] Without firm protections in place, each and every one of us—and our children and the people we love—is one hack away from a manufactured scandal, a deep fake, and AI-generated pornography.

---

44   Julie Jargon, "Fake nudes of real students cause an uproar at a New Jersey High School," *Wall Street Journal*, 2023, https://www.wsj.com/tech/fake-nudes-of-real-students-cause-an-uproar-at-a-new-jersey-high-school-df10f1bb.

Equality

I define equality in the context of TAPER as guaranteeing that all individuals have equal rights and opportunities in the digital realm, irrespective of their background or abilities.

Any solution to how we care for and use our data must have equality.

In the digital economy, we should all be equal. My digital "vote" should be equal to that of someone who lives in circumstances far removed from my comfortable American life. Every human should be given the opportunity to participate in this economy of the future.

Like many, especially in the tech world, I am also concerned about the people who don't have access to the internet, who don't have access to technology in their schools. We all need to share in this digital commerce—in what our data is worth.

In addition, if I might pause for a moment to point out that we are all interconnected. I can communicate with someone across the planet—and for the time we interact online, the world becomes a smaller place. I firmly believe that virtue does not end at our borders and everyone beyond is not an enemy. I think if we can harness the power of data and the paradigm shift I am suggesting, the world will be smaller and safer. We, as human beings, are less likely to begin a war with another country if we can see their citizens are not so different from us.

Ultimately, though, equality is about equal opportunity for every individual in the digital age to participate and to be protected. Everyone, not only the few.

Respect

In our TAPER concept, respect is upholding the dignity, privacy, and rights of individuals in all data-related processes.

I think for me, respect, in this context, is restoring our humanity. Instead of treating us like marionette puppets and pulling this string or that one to influence our decisions, we will regain our control.

We will be in the driver's seat, not the digital overlords.

Think of when we feel disrespected because someone talks down to us (or "mansplains" or is dismissive). I think these digital despots disrespect us by telling they only take this little bit of digital DNA from us, without admitting they will mate that data with other data in a "clean room."

They know who you are.

Remember our mantra from earlier in the book? Yes, your PII might be hashed into a long series of numbers and letters instead of your name. But it's just a game (or nonsense). These commercial entities gathering data pretend they don't know who we are—but just like the profiles of Erica, Jonathan, Rose, Kara, and Tom, it's amazing how much an entity can know about you—they know what you ate for breakfast, know you have four kids and their approximate ages, know what TV show you view on your Smart TV at 8:00 a.m., know where you are driving via your GPS, know where you travel to each year on vacation, and so on.

Most of us would be horrified to discover our house was bugged, our car had a transponder on it tracking our movements, and someone was watching us as we shopped, as we dined out—and knew what we watched on our TVs, and even where we are in our house as evidenced by our fitness tracker watch. Stalking us. But we are being digitally stalked—and it shows a basic lack of respect for us and our autonomy and freedoms.

We are being objectified completely. Not only that, we are being objectified for the purpose of economics—money. Our privacy is being invaded—to sell more cereal or to get us to buy a Bahamas package deal to get away in February. We are nothing more than aggregated data.

CHAPTER RECAP

We have arrived, in this first chapter of part II, at a solution-focused approach to data privacy. This chapter expanded beyond the C3s we discussed in part I—*consent, control,* and *currency*—to TAPER. The C3s and TAPER work together to form the foundational principles of data independence.

Once again, TAPER stands for:

Transparency (T)*:* Ensuring clear and open practices in data usage, storage, and sharing.

Accountability (A)*:* Entities handling data must be accountable for their actions and compliance with these principles.

Protection (P)*:* Strong measures to safeguard data against unauthorized access and breaches.

Equality (E)*:* Guaranteeing that all individuals have equal rights and opportunities in the digital realm, irrespective of their background or abilities.

Respect (R)*:* Upholding the dignity, privacy, and rights of individuals in all data-related processes.

Now that we have explored the principles needed to arrive at any true solution to the data dilemma we have and our inherent lack of privacy, we will go into detail on the *PDVK.*

It will be vital to our data independence!

CHAPTER 6

The Personal Data Vault Key: What It Does and How It Works

I am a firm believer that transparency goes hand in hand with collective intelligence.

—RANA EL KALIOUBY

Our PDVK is a paradigm shift of how we can take back the consent, control, and currency of our personal data. To be clear, this is a real, actionable solution. Too often, there is a sense of hopelessness (promoted, no doubt, by the Big Five digital overlords) that there is nothing that can be done. As someone who created algorithms for NASA, I know if we can do *that* and any number of amazing feats of human ingenuity, the only thing holding us back from solving this is our lack of resolve.

Before we look at how the PDVK will work, let's examine how our data is collected now.

THE CURRENT LANDSCAPE

Figure 6.1 is a single illustration of how the current digital landscape works when it comes to our data. This diagram is not meant to depict all data capture processes. I have taken one of our profiles and painted a typical scenario of how she is digitally surveilled—all while simply going about her everyday life.

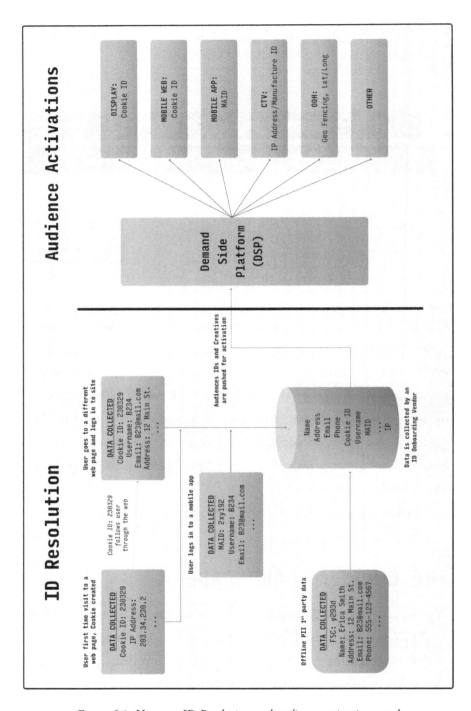

Figure 6.1: How our ID Resolution and audience activations work

OUR UNSUSPECTING MOM

Currently, Erica is a loyal Kroger shopper. She is pleased when she gets to the checkout to utilize the app and to receive coupons/discounts on the products she buys each week: name-brand cereal, her preschooler's favorite cookies, her husband's gluten-free bread, and her usual paper towel brand.

To receive these discounts, when the new Kroger's opened about three miles from her house, she signed up for a frequent shopper card (FSC). In fact, she also has cards with other retailers as she watches their sales to be sure to get the best prices for her large family. When she signed up, she was required to provide her PII—which in this case was her first name, last name, address, phone number, and email address (bottom left of the illustration).

Meanwhile, Erica is an avid online reader. She has subscriptions to the local newspaper in her suburb, as well as the *New York Times,* the *New Yorker,* and the *Atlantic.* She also avidly reads articles on film and politics, including visiting sites like the BBC, *Washington Post,* the *Wall Street Journal,* and more. Additionally, she has a hobby that requires art supplies, so she regularly shops online for that. The first time Erica visits any website (top right of the left-hand side of the graphic), a cookie is created. This cookie is unique to her and her browser. It can collect information such as domain, the content she is viewing, and other identifiers, such as IP address.

As she logs onto the internet in her home office and visits her favorite sites—stopping to go ahead and purchase the pants she has in her shopping bag at her favorite casual clothes retailer *J. Jill,* the cookie is still present and collects more information. For example, when she logs into her online magazine subscription, it collects her email and username—and is aware of which articles she is reading.

Next, companies that provide identity resolution services source digital information about Erica from various partners, using different methods, and then link this back to *offline* data. Thus, in our case example, Erica is linked via her email address to her FSC—which also links her to her online activity because it's all being stitched together.

The next day, Erica is on her mobile sports app to follow her favorite major league baseball team. The mobile app, which is a paid app, requires her username and password—she already gave them her email when she signed up. This data is now captured by the app—and unbeknownst to her is sourced to an ID resolution company. The ID vendor can again link Erica to her email address, only now they can add her MAID (mobile app ID) to her overall ID profile.

So why all these efforts? Because now advertisers across a wide variety of industries and sectors can reach her at various endpoints.

For example, a potato chip company whose brand she has long bought is launching a new flavor. She is their ideal target customer (well, not her but two of her kids and her husband, who all love chips). Consequently, the chip company would like to display their ads to her via multiple channels.

Using her cookie ID, when she loads up her favorite news channel, the first ad for the new chips pops up on the side. She looks at it but does not click on it. When she logs on to another news site, the ad is now a banner. When she goes on her Instagram account, one of the ads in her feed is a clever digital ad that is a "mockumentary" on chips. She clicks on it and laughs.

On her phone, she uses her GPS app to route her to Trader Joe's—she needs her GPS as it's a new location she is not familiar with. On the way, she sees—you guessed it—chip ads. This is because, using her MAID, these digital advertisers can even tell where she is geographically and place out-of-home (OOH) ads that are relevant to her geo fence.

Erica goes about her day until it's time to collect all the kids from their various lessons and sports, order take-out, answer work emails, supervise homework, and do the "hand-off" for bath- and bedtime to her husband so she can work on a project in her home office. At 9:00 p.m., at last, she and her husband sit down to talk about their days, the kids, the upcoming weekend plans, and to watch a television show they enjoy together.

As the police procedural show airs, an ad comes on for the chips. They are reaching Erica on her connected television because of the IP address, which was sourced from her digital web browsing. The brand has a comprehensive omnichannel campaign.

Is it any wonder she buys the chips?

THE PERSONAL DATA VAULT KEY (PDVK): TAKING BACK THE C3

There is a better way.

As I said earlier, if we can land a rover on Mars, we can solve the issues of data privacy. I firmly believe that what I am proposing can transform our world (something we'll discuss in the next chapter). It is a philosophical change, taking back control of our digital DNA.

Figure 6.2 shows a diagram of how the PDVK could work. I will walk you through the diagram so you have a full view of how this deterministic model could change all our lives. What I hope for this book to accomplish is to launch a conversation on how we can take back the C3 for ourselves.

First, each person can choose to have a verified PDVK if they want to participate in this new, secure system.

The other important player will be the data brokers. As I wrote in the previous chapter, brokers will need to have the infrastructure and systems in place, like Fidelity, to use an analogy, or other investment

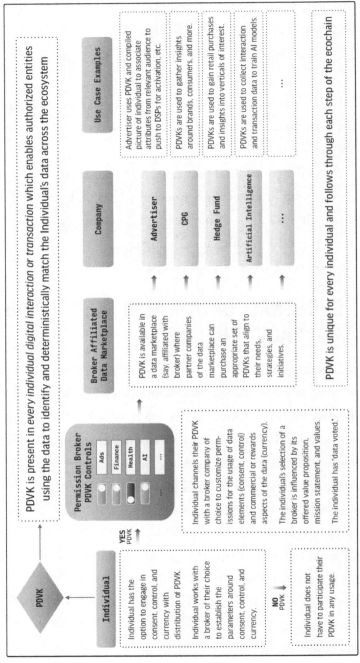

*Figure 6.2: A diagram of the PDVK—how it could
change the way we protect and use our data*

or banking firms, to guarantee, as much as possible, the security of your data. Your Uncle Harold cannot decide to become a data broker and set up a bunch of equipment in his basement as a side hustle.

The individual—such as our profiles of Erica, Jonathan, Rose, Kara, and Tom—decides which broker to use. Those brokers will be *competing* for your data. As such, they will need to work hard to embody the values they espouse, to provide transparency, accountability, protection, equality, and respect.

In short, in our digital future, you will vote with your "data vote" for those brokers who are going to care for your digital DNA in the manner you choose. Let's give three examples:

- Broker "Green": This broker espouses principles of sustainability. They have devised a strong pledge that companies to whom they give data to must sign and adhere to—or be removed from their brokerage house.

- Broker "Medical Breakthroughs": This broker offers data to advance medicine and studies, in order for scientists to conduct longitudinal studies.

- Broker "Show Me the Money": This broker seeks to maximize the monetary return for your data.

In keeping with TAPER, this system will have transparency. Because users have consented, there is no stitching profiles together. There is none of this silly "game" we have been playing with our data where "they" pretend they don't know who we are just because with have a PII.

But now the individuals are in charge. They are voting with their data. There will be people who will want to share *none* of their data— and that is OK too. For those who choose to share, they will consent to what to share, for what purpose, and for what monetization. The individual is respected based on the selections they opted for via the broker they chose.

In our diagram, you will also see four use cases (advertiser, CPG, hedge fund, AI). We will look at them in detail in chapter 8.

In addition, the PDVK could be a means when *you* or an advertiser, media source, or online commenter puts something onto the internet, we know who/how it originates—which as you will see in the next chapter could eliminate most sources of internet fraud.

Most importantly with the PDVK model, the individual is in control. They determine their privacy levels. We will at last have data independence.

CHAPTER RECAP

In this chapter, we at last lay out the details of your *PDVK.*

The principles of PDVK are rooted in the consent, control, and currency (C3) foundational trio we discussed in part I, along with the five principles of the PDVK, which are described through the acronym TAPER.

In order to see the difference and to understand how this shifts the paradigm, we first looked at how things are today—where what we do in our lives, both online and "in real life," are surveilled by data overlords. In our first graphic in the chapter, you can see how digital DNA is followed by corporations who collect our information.

The PDVK emphasizes the individual's power in selecting a broker based on value propositions, mission statements, and alignment with the individual's worldview. The overall theme is the empowerment of individuals in managing their personal data through PDVKs in a data marketplace environment.

These PDVKs are unique to *every individual* and are integral to *every digital interaction or transaction*, enabling authorized entities to deterministically match an individual's data across ecosystem.

In our next chapter, we will look at the many issues concerning the internet and data today—and the ways in which the PDVK will solve them.

CHAPTER 7

Before and After the PDVK: Use Case Examples

But we live in a complex world where you're going to have to have a level of security greater than you did back in the olden days, if you will. And our laws and our interpretation of the Constitution, I think, have to change.

—MICHAEL BLOOMBERG

As we saw in our last chapter, the PDVK will fundamentally change our relationship with our data. I presented examples of advertising agencies, CPG, hedge funds, and AI use cases in figure 6.2.

What can our use cases show us in terms of the before and after of this paradigm? How will the way we are tracked change? This chapter will then be a preview of chapter 8, which will explain how the PDVK will transform our freedoms and privacy and protect us from some of the darkest aspects of data in our lives.

ADVERTISING AGENCIES

Today. Advertising agencies serve as experts in targeting, planning, and achieving outcomes. At the heart of their expertise lies the relentless pursuit of data and the utilization of sophisticated analytics methods to extract actionable insights. The use of cutting-edge technology and data-driven decision-making processes ensures a competitive edge for their clients in an ever-evolving landscape of challenges and opportunities. Advertising underscores the central role of data and analytics in shaping outcomes and achieving objectives in today's complex world.

I would never want it to seem as if the ad agencies are the "bad guys" in this new digital economy. As I have said elsewhere in the book, those of us at the forefront of these changes in data, algorithms, and ads aspired to transform tasks most of us take for granted these days, like receiving coupons for products and brands we enjoy and can use.

Remember that ad agencies are simply serving the brands that hire them. Their livelihoods depend on results—do the consumers they want for their brand see the ads they place and then act on them?

In the realm of advertising, agencies serve as masters of data who have also mastered advanced analytics techniques to construct comprehensive portraits of individual consumers—those stitched-together 360-degree clones of us that we weren't even aware were being constructed, like digital Frankensteins.

At the core of their operations lies the sophisticated utilization of data science methodologies, enabling agencies to sift through vast amounts of information and distill actionable insights. By leveraging advanced algorithms and ML models, funnels, and the like, they can identify patterns, trends, and correlations within the data, allowing them to tailor messaging, content, and context with great precision.

This entire book has emphasized . . . *they know who you are.*

But the reality is they know who you are because of data.

Advertising agencies meticulously collect and analyze data from various sources to gain insights into consumer behavior, preferences, and interests. Advertising agencies excel in decoding the intricacies of consumer psychology. Through the strategic integration of demographic, behavioral, psychographic data, and many other data sources, agencies can construct highly detailed profiles of individual consumers, enabling them to craft personalized marketing campaigns that resonate on a deeply personal level.

Think of those "Hallmark" commercials, or holiday ads, that tug on the powerful emotional marionette strings of loneliness, nostalgia, grief, sensitivity, love, and family. We respond and choose products over and over because they resonate with us. For that matter, it does not have to be the deep pangs of a grocery chain's Thanksgiving ads. We respond to this sports drink because our favorite baseball player drinks it, or that chips brand because the mascot makes us laugh.

Furthermore, agencies are adept at orchestrating multifaceted campaigns that span across various channels and touchpoints, ensuring maximum impact and engagement. By harnessing the power of data-driven decision-making processes, they can optimize campaign performance in real time, adjusting messaging and content to align with evolving consumer preferences and market dynamics. The advertising world understands that the role of data and analytics is pivotal in shaping consumer experiences and driving business outcomes in today's hyper-connected world.

What about the after? In the envisioned new world of C3, where individuals possess full control over their personal data through their uniquely held key—their PDVK—the landscape of advertising undergoes a transformative shift toward a more deterministic model. In this scenario, keys to individuals' data must be actively sought and assembled by advertising agencies from the market, reflecting the explicit preferences and consents dictated by each person.

The individual *determines* what personal data they will release to those who wish to tailor ads to their buying and personal habits.

This change heralds a paradigm where the conventional approach to advertising is replaced by a highly respectful and targeted model, following TAPER. Here, advertisements are only delivered to individuals who have expressed a direct interest in or alignment with the products, services, or values of a company. This alignment is determined through the permissions individuals set through their agent/broker, effectively granting or denying access to their data for specific types of advertising.

The shift toward deterministic advertising underscores a mutual respect between consumers and advertisers. Individuals are no longer subjects of intrusive data mining without their understanding of what they "consented" to. Instead, they engage in a transparent exchange where they can choose to share their data with entities that respect their preferences and values, that respect their C3. This model empowers consumers to curate their digital experiences, only interacting with brands and services that resonate with their personal choices and desires.

For advertising agencies, this new world demands a new approach. Success in this environment hinges on their ability to understand and align with consumer wishes genuinely, fostering trust and respect. In doing so, advertising becomes not just a means to promote products but a service that adds value to the consumer's life by connecting them with solutions that fit their specific needs and preferences. The deterministic nature of advertising in this new world ensures a higher level of relevance and engagement, as individuals are more likely to be interested in the advertisements they see. It also builds brand loyalty and trust—the very things the advertising agencies spend a lot of time and money trying to figure out how to do.

Finally, this model not only enhances the effectiveness of advertising campaigns but also contributes to a healthier digital ecosystem where respect for personal data is paramount.

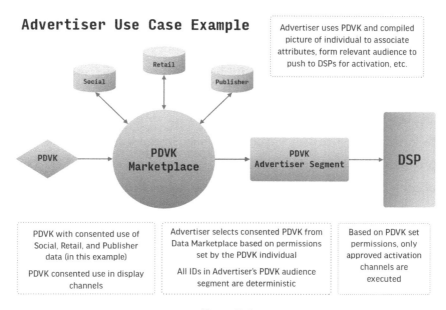

Advertiser Use Case Example

Advertiser uses PDVK and compiled picture of individual to associate attributes, form relevant audience to push to DSPs for activation, etc.

PDVK with consented use of Social, Retail, and Publisher data (in this example)

PDVK consented use in display channels

Advertiser selects consented PDVK from Data Marketplace based on permissions set by the PDVK individual

All IDs in Advertiser's PDVK audience segment are deterministic

Based on PDVK set permissions, only approved activation channels are executed

Figure 7.1

CONSUMER PACKAGED GOODS USE CASE

Today. In the competitive landscape of CPG, companies are constantly seeking innovative ways to have their brands reach their target audience effectively—one use case example. To reach an audience of interest, one such method is through the use of data from retailer loyalty programs, where shoppers sign up to receive discounts and coupons (something we saw Erica do in her profile). During sign-up, consumers typically provide personal information, such as their names, addresses, phone numbers, and email addresses. This information is an entry point into the vast world of data-driven marketing.

From this data entry point captures transactions and purchases at retailer properties. While consumers engage in online activities, cookies track their browsing habits. These cookies, which are unique to the user and browser, can collect a wide range of data, including domain, content viewed, and IP addresses.

Next, this digital footprint becomes more detailed as users interact with various online platforms, signing into different sites with usernames and emails. Furthermore, identity resolution companies specialize in piecing together these disparate pieces of digital information. They aggregate data from various online and offline sources, linking them to form comprehensive consumer profiles. In the scenario of a loyalty program participant, their email address becomes a key identifier, connecting their in-store activities with their online behaviors.

When consumers use mobile apps, additional identifiers, such as MAIDs, are captured and incorporated into their profiles. These enriched profiles enable CPG companies to execute omnichannel marketing campaigns tailored to the consumer's interests and behaviors. For example, a leading potato chips brand launching a new flavor can now target frequent snack enthusiasts through multiple channels. By utilizing cookie IDs, the brand can place personalized offers on web displays and mobile websites. The use of MAIDs allows for reaching consumers through mobile apps. Moreover, by analyzing geographic data, OOH advertising can be optimized within the vicinity of consumers' most frequent locations. Connected TVs provide yet another avenue, leveraging IP addresses from digital browsing to serve targeted ads during viewers' favorite shows.

Such data-driven strategies enable CPG companies to not only advertise but also gain insights into product performance across different channels and touchpoints. This informs future product development, pricing strategies, inventory distribution, and overall marketing effectiveness.

What about the after? What happens in the new data landscape shaped by C3, TAPER, and PDVK frameworks? Here, the narrative of consumer data sovereignty takes center stage. Here, individuals hold the reins of their personal data through secure digital keys. These keys represent their consent, control, and currency in the data ecosystem, allowing each person to dictate who accesses their personal information and the purposes for which it is used.

Consumers, when participating in loyalty programs like FSCs, are no longer passively providing personal information; they're entering into transparent data partnerships. These partnerships stem from the marketplace of PDVKs—brokered exchanges where individuals consciously decide which aspects of their data, such as purchase history or product preferences, they are willing to share and under what terms.

CPG companies and retailers must now engage with customers on the grounds of transparency, navigating a market where data access is granted by the individual. This transition has significantly altered the advertising landscape. Promotional materials are now a tailored experience, designed to align with individual interests and the stipulations set within their PDVKs.

The concept of a data footprint—encompassing online site visits, app usage, offline data, and other data—is securely guarded. Each touchpoint of an omnichannel campaign, from web displays to mobile apps and connected TVs, requires the consumer's digital key (PDVK) for access. This ensures that engagement with brands is consistent with each individual's C3 parameters, fostering a mutually beneficial (*currency* from C3) and respectful (*respect* from TAPER) relationship.

This new paradigm elevates the consumer's role from a passive subject of marketing campaigns to an active, empowered participant in the data economy. It allows each person to manage their digital identity and personal information with an unprecedented level of control.

Transparency and respect are the cornerstones of the transformed relationship between CPG companies, retailers, and consumers, leading to a market that rewards those who understand and honor this new model of data partnership. In this reimagined ecosystem, choice and trust are the standard. And individuals are in the driver's seat.

CPG Use Case Example

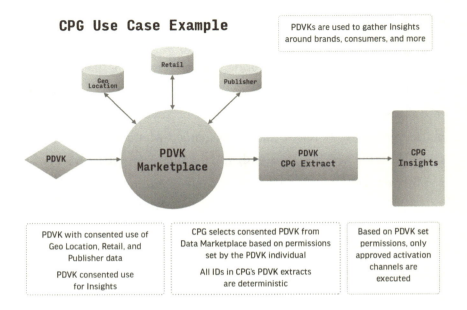

PDVKs are used to gather Insights around brands, consumers, and more

PDVK with consented use of Geo Location, Retail, and Publisher data

PDVK consented use for Insights

CPG selects consented PDVK from Data Marketplace based on permissions set by the PDVK individual

All IDs in CPG's PDVK extracts are deterministic

Based on PDVK set permissions, only approved activation channels are executed

Figure 7.2

HEDGE FUND USE CASE

Today. In the midst of the COVID-19 pandemic, Link Logic Capital, a NYC-based hedge fund, developed a unique strategy to decode a picture of supply chain disruptions. Recognizing the invaluable insights hidden within retailers' shopping data, the fund embarked on a quest to attain these data assets and untangle the supply chain complexities during the pandemic. The eccentric founder and CEO of Link Logic Capital has a fascination with data and penchant for unearthing unconventional investment strategies that knows no bounds. Armed with a team of data scientists, the CEO sets out to harness the power of retail data in decoding the supply chain puzzle.

As the pandemic rages on, the CEO and his team meticulously gather data from a vast array of retailers spanning the nation. From grocery giants to specialty stores catering to appliances to construction materials suppliers and many more retailers, every dataset holds a piece of the supply chain saga. His data scientists are seasoned experts and understand the synergy of diverse data sources. They use data fusion techniques over these diverse datasets and other datasets into a "super dataset." Using sophisticated ML algorithms, they transform this dataset into actionable insights, painting a clear picture of supply chain, at the nation, state, county, and zip code levels.

As the CEO and the company's traders delve deeper into the insights, patterns and understandings begin to emerge. Week-over-week fluctuations, year-over-year trends, anomalies in sales data, dwindling sales of certain items, fluctuations in inventory levels, etc., reveal subtle clues about the ebb and flow of supply chains across industries. These insights are unmatched and are used to make investment decisions. Amid the chaos of COVID-19, Link Logic Capital achieves unprecedented returns on investments with the unique clarity they had. The Link Logic Capital team were armed with data-driven foresight that no one else had on Wall Street.

What about the after? In this reimagined world where data sovereignty is supreme, the story of Link Logic Capital unfolds differently. Unlike the data gathering of the past, the hedge fund now operates within a new paradigm where the consent, control, and currency (C3) of data are governed by each individual through their PDVK. These keys are not freely available but are entrusted to entities by individuals based on personal decisions regarding data usage and privacy preferences. The hedge fund's quest to decode the intricacies of supply chain disruptions during the COVID-19 pandemic follows a different process: gaining access to retail data now requires navigating a marketplace of individual data keys.

Link Logic Capital can no longer rely on a direct and unfettered pipeline of data from retailers. Instead, they must seek the

PDVK market for such individual data. The value proposition has to attract needed PDVK to realize the use case. Retailers still play a crucial role in this new ecosystem, for example, they can share nonpersonalized data like inventory levels without infringing on individual privacy. However, for more granular insights that can truly illuminate supply chain dynamics, Link Logic Capital must engage with individuals through the PDVK market, seeking those who are willing to share their data under clearly defined terms that fits every PDVK's C3.

The marketplace for individual data keys is competitive and requires that Link Logic Capital not only offer a compelling reason for individuals to share their data but also adhere to the highest standards of data privacy and ethical use—following TAPER. Each PDVK transaction is a partnership, with individuals exercising their "data vote" to support the fund's research while retaining control over their personal information and realizing currency.

As Link Logic Capital navigates this new data landscape, it finds that its success hinges not just on its analytical capabilities but also on its ability to build trust and respect among a network of individual data partners. The hedge fund's journey becomes one of collaboration and mutual benefit, reflecting a broader shift toward a more ethical and sustainable model of data utilization. This realization is not unique to Link Logic Capital but to all companies and entities seeking our individual data. In this new era of data sovereignty, Link Logic Capital emerges not just as a hedge fund with a keen eye for investment opportunities but as a pioneer in ethical data practices, setting a standard for how industries can and should engage with individuals in the collection and use of data.

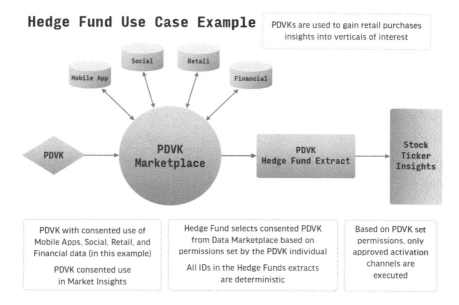

Figure 7.3

ARTIFICIAL INTELLIGENCE USE CASE

Today. Mind Read Insights AI, aka MRI AI, is a pioneering AI firm headquartered in Silicon Valley. It developed a revolutionary approach to decipher the intricacies of human intent using AI. Leveraging a myriad of behavioral, sentiment, psychological, transactional, and purchase histories; social media posts; online searches, and many (*many!*) more data sources, MRI AI sets out to unveil the hidden desires and motivations that drive consumer actions.

The company's acronym is telling by itself! And, it is also how they describe themselves: "We believe that every data point is a thread in the intricate tapestry of human intent. We use AI to obtain a psychological MRI for every consumer." At the helm of the company stands its CEO, who holds a Ph.D. from a top US university and is a luminary in the realm of cognitive psychology and AI. With a team of expert data scientists and behavioral psychologists by her side, she embarked on a quest to create the ultimate tool for marketers and companies: IntentGPT.

IntentGPT was developed tirelessly to distill this wealth of information into actionable insights. Through the fusion of these (available and attainable) data sources and system training over these data assets, IntentGPT emerged as a capability unparalleled by any other in the marketplace, unearthing unprecedented capabilities in consumer behavior. With IntentGPT in hand, marketers and companies gained unprecedented access to the inner workings of the consumer mind. The possibilities were limitless, from predicting future purchase behaviors to crafting personalized messages and developing marketing strategies, etc., businesses are able to ensure an almost-perfect score, almost 100 percent, of any desired outcome!

These capabilities are now more powerful than ever before—they are AI powerful. A natural question you might ask is, Are we "really" free, and what is free will then? Can MRI AI control every individual? Are we mere consumers being manipulated with those puppet strings to act in a certain way knowing that an entity can not only "read our mind" but "program our minds" to ensure the desired outcome? The new paradigm of C3, TAPER, and PDVK addresses these questions and empowers every individual to make their own (truly free!) decisions.

What about the after? In a world increasingly vigilant about personal data rights and privacy, where C3/TAPER/PDVK reign, the narrative of MRI AI unfolds as a cautionary tale. Founded with the ambition to revolutionize marketing through the power of AI, MRI AI sought to delve into the deepest corners of human intent. Their flagship product, IntentGPT, promised unprecedented insights into consumer behavior. In the new world, the data had to be acquired in a C3 and TAPER world. MRI AI needs vast datasets that span across behavioral, psychological, transactional dimensions, and more.

However, as the mission and foreseen capabilities of MRI AI are widely known (transparency) in the new world of TAPER, a global discourse emerged, centering on the ethics and implications of such profound insights into consumer behavior. The notion of an *entity* possessing the ability to not just understand but potentially manipulate the very fabric of human decision-making sparked a fierce debate. Critics warned of a dystopian future where free will was undermined by AI, a world where "gaslighting on steroids" was not just a possibility but a reality. The power of MRI AI and other such use cases, while mesmerizing, began to be seen as a double-edged sword, with the potential for misuse overshadowing its benefits.

With TAPER, as awareness grew about such technology use cases, so did the resistance. A movement, powered by the will of each individual, began to assert their "data vote" via their consent and control of their PDVK. The vast majority opted out, choosing not to contribute their data to entities perceived as too intrusive or wielding too much power over individual autonomy. This collective action was not just a statement about privacy but a stand for human dignity and the right to an inviolable mental space. Currency offered by such technology firms was not sufficient to collect the needed data assets at scale. Without the scale of data needed to power its engines, MRI AI found itself in an untenable position. The AI, which thrives on the breadth and depth of data to function effectively, was starved of its primary resource. The market for their services never took off the ground, not because the technology was flawed but because the societal consensus deemed the cost to personal autonomy too high a price to pay.

MRI AI folded—a testament to the power of collective action in the digital age, via the "data vote" enabled by C3/TAPER/PDVK. The collapse of such a company and technologies serves as a stark reminder that the path forward for AI is one that must be treaded carefully, with ethical considerations and respect for individual autonomy at its core. The narrative of MRI AI is not just a story of technological hubris but a cautionary tale about the importance of balancing innovation with the intrinsic values of human dignity and freedom. Only our "data vote," via our data independence in a world where C3/TAPER/PDVK reigns, can protect our freedom.

CHAPTER RECAP

If we embrace a new paradigm for how our data is handled, then the use cases explored in this chapter will transform the way we interact with those who currently exploit our data.

The use cases demonstrate a whole new freedom—where the data is *ours* and we are in the driver's seat, not the corporate entities and others who collect our digital DNA.

In fact, in chapter 8, we will explore how this PDVK can address some of the biggest data dilemmas of our time.

How the PDVK Can Solve the Big Data and Privacy Issues of Our Time

Fake News, Fraud, and More

*In the digital era,
privacy must be a priority.
Is it just me, or is secret blanket
surveillance obscenely outrageous?*

—AL GORE

From the dawn of humanity until 1954, no one ever ran a four-minute mile. It was impossible. Humans just weren't made to run that fast—not even elite athletes.

Or so was the entrenched belief.

Along came Roger Bannister in 1954. He broke the four-minute time. He also broke the paradigm.

As of this writing, nearly 1,800 athletes have run a four-minute mile or better, and the current world record holder ran it in under 3.5.

Elite runners were freed from the paradigm that such a barrier existed.

We all need a data paradigm shift—two of them. First is that the problem is too big to solve. "Impossible!" Our leaders on Capitol Hill are generally not from the world of tech—or data privacy. They may or may not have the staff to delve into the whole topic.

The following is an actual quote from a congressional leader to Mark Zuckerberg.

> "What if I don't want to receive [ads for chocolate]?"[45] Bill Nelson, from Florida, apparently had an issue with chocolate ads. Zuckerberg instructed Nelson that users could turn off third-party information within Facebook if they don't want that info used to select the ads they see. But, Zuckerberg added, "Even though some people don't like ads, people *really* don't like ads that aren't relevant."[46]

Note how "relevant ads" is the dangling reward. Not currency.

But we need to have a conversation on digital privacy and these rights—now. Because these problems will not go away in the future.

Which leads me to the other paradigm shift. We the people need to start thinking of our data as *ours*. We need to think of it as an asset. As having value—and that asset belongs to us.

Yet, solving this issue with legislation alone is untenable—because with a problem this vast and woven through all facets of our lives, we must have a solution across our country and ideally the globe and not piecemeal legislation state by state. California's issues related to data privacy and security do not end at her borders.

As I considered what the PDVK could do, I realized that every problem and issue I raised in this book at least *could* be solved with this solution. The legal particulars, the administration and operation,

---

45 Ibid.

46 Ibid.

etc., of this can be up to the great minds—but again, we need ethicists and privacy scholars, and ordinary citizens, to be part of the discussion and not leave it only to the digital overlords, because that just keeps us as digital serfs.

Let's examine some of the bigger data and data privacy issues of our lives—knowing that as soon as this book is printed, there will be new stories of data breaches, or clever hackers, of legal loopholes, and things I cannot even predict of now (though I have some guesses).

FAKE NEWS

Fake news came into the lexicon in 2016—and perhaps not how you remember it. According to the BBC, Buzzfeed's editor noticed stories—crazy, fake stories—originating in Macedonia, and spreading from there.[47] One example headline was "FBI Agent Suspected in Hillary Email Leaks Found Dead in Apparent Murder-Suicide."[48] Buzzfeed ended up tracking down 140 news sites disseminating fake news about the American election.[49]

As time has passed, however, politicians (and plenty of others) have weaponized "fake news"—by forwarding actually fake news and false stories on their social media feeds, or even in speeches. In addition, it is weaponized by denouncing any *true* but unflattering portrayal as "fake news," thus causing ordinary citizens of all political stripes to doubt what is true.

It is positively Orwellian, as this excerpt from *1984* depicts:

---

47 Mike Wendling, "The (almost) complete history of 'fake news,'" BBC, https://www. bbc.com/news/blogs-trending-42724320.

48 Ibid.

49 Ibid.

> The party told you to reject the evidence of your eyes and ears. It was their final, most essential command. His heart sank as he thought of the enormous power arrayed against him, the ease with which any Party intellectual would overthrow him in debate, the subtle arguments which he would not be able to understand, much less answer. And yet he was in the right! They were wrong and he was right.[50]

But the PDVK could eliminate the ability to manipulate us. Anything shared on the internet would either bear the unique PDVK of the original poster or be unverified and therefore a suspect source. We would know where it originated—and by whom.

Certainly, knowing a medical article originated from the *Journal of the American Medical Society* would be a more trusted source that an article that *looks* suspiciously like a *JAMA* article but was actually written by Stefan who operates a 'bot farm out of his mother's basement in Macedonia.

According to research from the University of Colorado Boulder, Facebook is the most popular conduit for misinformation and fake news.[51] Additionally, the people at each end of the extremes—those who identify as extremely conservative (a 7 on a scale of 1–7) or extremely liberal (a 1 on the conservative scale) account for the biggest volume of misinformation or fake content shared, with conservatives

---

50   George Orwell, *1984* (New York, NY: Signet Classics, 1961).

51   Lisa Marshall, "Who shares the most fake news? New study sheds light," 2020, https://www.colorado.edu/today/2020/06/17/who-shares-most-fake-news-new-study-sheds-light.

edging out liberals 26 percent to 17.5 percent as far as disseminating the most fake news and images on Facebook.

The researchers, Toby Hopp and colleagues, published their finding in the journal *Human Communication Research*.[52] He said, in an interview, "Despite the fact that we tend to call it 'fake' news, a lot of this stuff is not completely false. Rather, it is grossly biased, misleading and hyper-partisan, omitting important information."[53]

In fact, Hopp prefers the term "countermedia" rather than fake news. What is *not* fake is its corrosive effect on politics and our lives—both real life and online.

In fact, fake news, disinformation, and misinformation can be literally deadly. Marcia McNutt, president of the National Academy of Sciences of the United States, said (referring to the COVID-19 pandemic) in a joint statement of the National Academies posted on July 15, 2021: "Misinformation is worse than an epidemic: It spreads at the speed of light throughout the globe and can prove deadly when it reinforces misplaced personal bias against all trustworthy evidence."[54]

The PDVK will take away the power to mislead. Sure, you can put out disinformation that repeats conspiracies and falsehoods. But the PDVK will allow the world to know if it came from Stefan in Macedonia or the surgeon general.

---

52   Toby Hopp, Patrick Ferrucci, and Chris J. Vargo, "Why Do People Share Ideologically Extreme, False, and Misleading Content on Social Media? A Self-Report and Trace Data–Based Analysis of Countermedia Content Dissemination on Facebook and Twitter," *Human Communication Research* 46, no. 4 (October 2020): 357–384, https://academic.oup.com/hcr/article/46/4/357/5840447?guestAccessKey=e1548abf-a0ae-469a-98f5-a9a04b0b769e&login=false.

53   Marshall, "Who shares the most fake news? New study sheds light."

54   Esma Aimeur, Sabrine Amri, and Gilles Brassard, "Fake News, Disinformation, and Misinformation: A Review," *Social Network Analysis and Mining* 13, no. 1 (2023): 30. doi: 10.1007/s13278-023-01028-5.

DATA INDEPENDENCE

FRAUD

We all know fraud is a pervasive problem, but the numbers are staggering.

First, before I share them, I will suggest that the FTC is a natural starting point as an organization to tackle the problems of fraud and our data. It could become the Federal Data and Trade Commission (FDTC). The FTC is already involved with trying to eradicate fraud—which is a bit like playing Whack-A-Mole. Every time a type of fraud or scam becomes better known, scammers invent a new one. Meanwhile, the FTC has offered sobering numbers as far as fraud losses in 2022. The figure, almost $8.8 billion, is a 30 percent increase over the year before.[55] The FBI's Internet Crime Complaint Center (IC3) reports the numbers in figure 8.1.

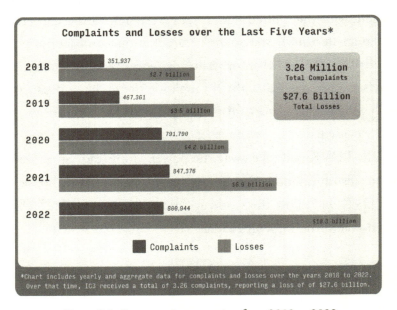

*Figure 8.1: Internet crime reporting from 2018 to 2022
based on complaints received from the Internet Crime Complaint Center (IC3)*

Lesley Fair, "FTC crunches the 2022 numbers. See where scammers continue to crunch consumers," 2023, https://www.ftc.gov/business-guidance/blog/2023/02/ftc-crunches-2022-numbers-see-where-scammers-continue-crunch-consumers#:~:text=Imposter%20scams%20top%20the%20Fraudulent%20Five.%20In%20terms,%234%20Investments%2C%20and%20%235%20Business%20and%20Job%20Opportunities.

With tools, technology, robocalls, bots, and AI, this figure will only climb. The 2022 FTC numbers suggest some disturbing trends about what scammers are up to.

Imposter scams top the "Fraudulent Five." In terms of the number of fraud reports received, the first one on the list is imposter scams, followed by online shopping; next is prizes, sweepstakes, and lotteries; number four is investments, and rounding out the fraudulent five are business and job opportunities.[56] With the sheer number of online dating sites and the people on them (for one example, Tinder has 75 *million* active monthly users),[57] the modern "lonely hearts" scams lure people in with a sob story, reeling it out little by little so as to not raise suspicion with requests for money. The dating scam can also be about an "opportunity," such as an investment, after building trust—and the catfish or fake profile pretends to just want their newfound friend to benefit. The lottery scams are often very real-looking emails declaring someone has won the lottery and all that is required is a "processing fee" and credit card number.

Losses to business imposters have skyrocketed. People's losses to business imposters topped $2.6 billion in 2022.[58] This can be anything from fraudulent links or rerouting to a malware site. This is very troubling because it diminishes consumer confidence.

Investment scams claim the highest losses to consumers. In 2022, consumers reported losing more money—$3.8 billion—to this kind of fraud than to any other category.[59]

---

56 Ibid.

57 Mansoor Iqbal, "Tinder revenue and usage statistics (2024)," 2024, https://www.businessofapps.com/data/tinder-statistics/.

58 FTC.

59 Ibid.

Scammers are on social media too, just like the rest of us. Losses facilitated via social media platforms totaled $1.2 billion in 2022.[60]

But the PDVK can help stamp out this kind of fraud—as well as help obtain justice when it is discovered. Consider this: years ago, when it was time to invest our money, we often went to a bank or other commercial building and met our broker or banker. We shook hands, we saw the office, we met the tellers, and so on. And while fraud can and does happen at both large and small "real" institutions, with a "real" presence, for fraudsters, that is certainly a bigger undertaking.

However, scammers create online businesses that look "just" like the real business, and we have to be incredibly careful about what we click on, even if it seems like our own financial institution. There are bot farms in far-flung locations, actively hacking, phishing, and more. Imagine how much more secure we would be if there was a verified PDVK associated with a website, a profile, a news source, those institutions that handle money and other transactions, etc. That we could know with confidence that there was a real entity or human at the other end of our transactions and interactions.

I think of people like our elderly citizen, Rose. She or a family member could check on the verification of the PDVK before handing over money. All of us, the digital citizens of the world, could rely on the PDVK before we fully trust an online merchant—or agree to provide our name and address or other information to an entity in the digital ecosystem.

We will discuss data breaches next, but some of the points I've raised are about the end users. Regular people being scammed. Often, however, acts of fraud are complex, international, the kinds of mass-scale crimes that are vigorously pursued by the authorities, but increasingly sophisticated operations, AI, and other technological advances

---

60   Ibid.

make the work of catching the fraudsters challenging. For example, one newer online fraud scheme is called the AnyDesk scam. In this con, an online ad offered an attractive apartment at an excellent rental price. The victims were promised more photos and a video of the property. However, in order to receive them, they needed to download the free AnyDesk remote desktop software. After the images were transferred, the fraudsters were able to maintain access to their victims' computers using the remote desktop software and eventually accessed personal details and bank accounts to steal money.[61]

Therefore, the PDVK will help defend innocent global citizens on a large scale against some of these common online fraud schemes that utilize our data in malicious ways. However, it will also help law enforcement and those organizations trying to eradicate fraud so we can enjoy our digital lives safe from scammers.

DATA BREACHES

No matter what we do, it is likely that data breaches will occur—hackers and bad actors are relentless in their pursuit. But protecting the PDVK will be imperative. When TAPER is implemented, it will determine who can "play" in the space of the digital universe as a verified digital citizen of entity.

However, with this new level of security (the PDVK) and requirements of data brokers, our data should be more secure than ever before.

Data breaches involve cybercriminals infiltrating "computer systems, networks, or databases to gain access to confidential information. Breached data can include personal information, financial records, intellectual property, or any other protected information

---

61   Douglas Jones, "Biggest scams of 2021, and what to watch out for
      in 2022," 2021, https://www.abc10.com/article/news/nation-world/
      scams-2021-watch-out-beware-2022/507-206e0640-a39a-4882-bf80-06905c2aa52c.

that falls into the wrong hands. The consequences of a data breach can be severe such as financial losses, reputational damage, legal implications and potential harm to victims. This can include anyone from individual consumers to smaller businesses and even large multinational enterprises."[62]

Now, very often the terms "data breach" and "cyberattack" are used interchangeably. However, a data breach generally focuses on "unauthorized access to the data. For example, a hacker gains access to users' names, Social Security numbers, and passwords."[63] On the other hand, cyberattacks are usually considered "malicious activities that cybercriminals use, such as malware infections and phishing schemes targeting computer systems."[64]

This can be done physically by accessing a computer or network to steal local files or by bypassing network security remotely—the latter method is the one usually utilized to target companies.[65]

Earlier in the book, I cited some of the biggest data breaches in history. Losses of hundreds of billions upon billions of account information and personal data have occurred around the world, costing an accumulated hundreds of billions of dollars. IBM, for example, cites the average global cost of a data breach is $4.5 million per company.[66] Some of the breaches have had astronomical costs. For example, the

62 Monique Danao and Kiran Aditham, "What is a data breach? definition, examples, and prevention," *Forbes*, 2023, https://www.forbes.com/advisor/business/what-is-data-breach/#Data%20Breach%20Definition.

63 Ibid.

64 Ibid.

65 "Data Breaches 101: How they happen, what gets stolen, and where it all goes," 2018, https://www.trendmicro.com/vinfo/us/security/news/cyber-attacks/data-breach-101.

66 "Cost of a data breach report 2023," IBM, 2023, https://www.ibm.com/reports/data-breach.

2017 Equifax breach cost the company $1.4 billion in fines, settlements, and other costs.[67]

The GDPR demands that companies must announce when they have been breached—and those breaches can incur fines of up to twenty million euros or 4 percent of a company's previous year's revenue. Breaches impact a company's bottom line, its reputation—and it impacts those whose sensitive data and privacy have also been breached.

When businesses are breached, compromised data may include personal information like names, addresses, and email addresses, Social Security numbers, birth dates, and other PIIs. When point-of-sale machines are breached, stolen data may be credit card transactions.

Then what happens to this breached material? It depends on the motivations of the cybercriminals. For most, it is greed. In the case of the Ashley Madison breach of 2015, hacktivists "stole and dumped 10GB worth of data on the Deep Web. This included the account details and personally identifiable information (PII) of some 32 million users, as well as credit card transactions."[68]

Medical and health care information has also been breached, a clear violation of HIPAA, of course. Anthem, the health department of Hong Kong, and other medical organizations have had breaches—including of medical data, but also the kind of nonmedical personal identification data hackers frequently seek for Deep Web marketplaces.[69]

The government is just as vulnerable to hackers. In one breach in 2015, hackers gained access to more than eighteen million federal employee records, including Social Security numbers, job assignments,

---

67   "What is a data breach?" IBM, https://www.ibm.com/topics/data-breach.

68   "Data Breaches 101: How They Happen, What Gets Stolen, and Where It All Goes," 2018, https://www.trendmicro.com/vinfo/hk/security/news/cyber-attacks/data-breach-101.

69   Ibid.

and training details.[70] Even credit card providers and banks, who should keep our data in a digital vault of sorts, have been subjected to breaches.

One of the ways in which the PDVK will help us is that the PDVK will ensure that entities on the internet that have keys have been verified in the digital ecosystem.

We will also be able to determine who has our personal data—and how much of it we are willing to share. We will have the control. Additionally, when we declare our data independence, what we are demanding of the world is protection of our digital privacy and digital rights.

When we discussed data brokers, for example, these entities will have to have extraordinary infrastructure and so on, in order to be deemed qualified to handle our digital lives. Banks and financial institutions must abide by certain rules. The SEC regulates the securities market. When we visit a doctor who is board-certified, we expect them to meet certain standards. This *new* digital ecosystem will need to bring together the whole of the country—and ideally eventually the globe—in order to strengthen the infrastructures and systems in place, and ensure our data is better protected.

CHAPTER RECAP

We spent much of the previous chapters discussing the ways in which our privacy has been violated. In addition, we all recognize the internet has the potential to be a great equalizer—to bring knowledge to us all, to be able to share ideas. Later, you will hear more about this vision—I have a dream of the digital citizens of the future.

---

70   Ibid.

In this chapter, we explored how the PDVK can solve some of the big issues that plague us right now—whether that is fake news, data breaches, fraud, phishing, or other pressing questions about how we can protect ourselves in the data ecosystem.

In our next chapter, we are going to revisit our characters—Erica, Jonathan, Rose, Kara, and Tom. Their lives are going to look pretty different with their PDVKs.

Then, let's see how the PDVK transforms their (and our) worlds in the future in part III.

CHAPTER 9

Our Profiles in a PDVK World: Finding Data Freedom

Freedom is something that dies unless it's used.

—HUNTER S. THOMPSON

In a world where C3, TAPER, and PDVK are the norm, Erica's morning unfolds differently.

As Erica enjoys her early morning tea, the choices on her Wi-Fi-enabled TV are no longer silently feeding into an extensive ID Graph. Thanks to the principles of C3, she has control over her data, deciding what information her TV shares. Her viewing choices, whether CNN or her kids' favorite cartoon shows, remain her private preference, not a data point for unsolicited profiling. Her privacy is further protected by the TAPER principles, enforcing her consent and control choices. For example, she opted that her shopping data from loyalty cards is no longer mixed with external sources to profile her household—that was her choice. She still enjoys her discounts, but now they come on her terms! She prefers to buy her Pop-Tarts, Lunchables, and dinosaur-shaped chicken nuggets without anyone trying to extrapolate how many children she has and their ages.

As Erica navigates her morning—dropping her kids at school, grabbing her favorite Starbucks drink, and shopping online—each digital interaction is a testament to her autonomy. The offers and advertisements she receives are tailored to her preferences, set and controlled by her, through her PDVK broker, respecting her choices of what to share and what not to share.

In the data-empowered world Erica inhabits, her daily digital interactions are not just passive experiences but active choices. By agreeing to share her geolocation data through her PDVK, Erica receives real-time offers that are precisely tailored to her preferences. This geo-sharing is a conscious decision, enabling location-based services to enhance her day-to-day activities with relevant, timely offers and information as she runs errands in her "mom bus"—the family's Toyota minivan.

As Erica sits in her camping chair at the sideline of her daughter's soccer game, she does a little work on her laptop, then hops online. Once combing the internet, using the hotspot on her phone, her online behavior, from shopping to minivan browsing (the family van's odometer just passed 200,000 miles), contributes to her digital footprint, but in a way that she controls.

How did this happen? By signing her PDVK with a broker that resonates with her values, Erica actively participates in the data ecosystem on her terms. In her case, she signed to a broker committed to sustainability. While that reflects her beliefs and values, this arrangement also ensures that she benefits from the monetization of her data too. It's a symbiotic relationship where her data contribution is reciprocated with value-added services and rewards. This nuanced approach to data sharing represents a shift from passive data collection to a more equitable, C3-based model where individual preferences and values are at the forefront of data exchange.

This shift in data dynamics illustrates a paradigm where personal data is not relentlessly exploited but is respected and valued. Businesses still gain insights from consumer behavior, but in a manner that respects individual privacy and choice. Erica's story in this new era is not just about the technology she uses but about the respect and control she experiences in her digital life.

In this new reality, Erica's digital interactions are still valuable for businesses, but they are no longer relentlessly mined and monetized without her input. Instead, there's a harmonious balance where her data serves her needs as much as it does for businesses, fostering a digital ecosystem built on mutual respect and benefit, including the currency aspect of C3.

In the transformed world where C3, TAPER, and PDVK are the norm, Erica's engagement with digital services takes on a new dimension of personalization and control. Her digital interactions, from shopping to entertainment, are now a reflection of her choices, safeguarded under the principles of C3 and TAPER. This change empowers Erica, allowing her to navigate the digital world in a way that respects her freedom of choices and echoes her individual values. Her story is a testament to how personalized control over data can reshape everyday experience, aligning technology with personal ethics and preferences.

Oh! And regarding the satellite images capturing the parking lot of the grocery store? They cannot serve the hedge funds, since there is no guarantee that all drivers agreed to share their locations. This ensures that location data isn't exploited without everyone's explicit consent.

01100010 01110010 01100101 01100001 01101011

In the era where C3, TAPER, and PDVK reign, Jonathan's life as a travel-savvy executive takes a turn toward data empowerment and sovereignty.

Jonathan, accustomed to a fast-paced lifestyle of business travel and high-end preferences, has become acutely aware of his data's value. He knows that his every online search, purchase, and digital interaction is a treasure trove for advertisers, brands, financial services, and more. In Jonathan's case, he is a high-end brand's dream consumer. He buys a new car every three to four years—always a BMW. He has a luxury watch collection, inspired by the Rolex he inherited from his grandfather as a teen, and now encompassing about twenty timepieces he has bought over the years. And then, of course, it's his travel—often traveling business class to cities like Zurich, Hong Kong, and Abu Dhabi, where he always stays at five-star hotels.

In this new world, however, he is no longer a passive source of data. Jonathan—like every individual—is now data independent, and he understands the mechanics of how companies profit from personal data. With this knowledge, he exercises his consent and control to share his PDVK with an agency that maximizes his data currency.

As an example, he learns about Retail Media Networks and grasps the concept that even mundane activities, like shopping at a retailer, are part of a larger scheme of data monetization. Jonathan sees an opportunity to maximize his currency aspects in every activity he takes. He chooses to share his data on his terms. He signs up with a broker that specializes in maximizing the value of personal data for individuals like him—people with high disposable income. Through C3 and his PDVK, Jonathan controls what data he shares, ensuring his privacy and preferences are respected. His broker ensures TAPER is implemented.

His data, spanning from his travel habits to his online shopping, his in-store shopping, his restaurant choices, his search and reading habits, his vital signs monitored through his smartwatch, and much more, is shared through his PDVK. However, this sharing is calculated, consented, and designed to maximize the currency value of his data—that was Jonatha's choice and freedom. Jonathan feels empowered; he's not just a high-earning executive but now also earns from his data—every second, every day.

While flying business class or browsing luxury cars online, Jonathan knows that the targeted ads he sees are there because he allowed them. He is no longer bombarded with irrelevant ads. Gone are the days when his social media feeds used algorithms that largely missed the mark. Instead, he receives offers that align with his lifestyle and interests, a direct result of his selective data sharing. Furthermore, his consent to share data with ride-share companies means that when he lands in LA, the offers he receives are curated to his preferences, maximizing both convenience and his data's profitability. Even his casual posts on social media or the magazines he reads online contribute to this new revenue stream. He is aware that his data is used with other companies spanning various industries (for example, insurance, banking, automotive, credit cards, etc.) as he wanted, to maximize his data currency returns.

In Jonathan's world, the concept of soft totalitarianism, where individuals are manipulated unknowingly through digital means, is a thing of the past. He is no longer a rat in a maze, lured by the pungent Stilton of entities seeking his data. Instead, he is a conscious participant, a savvy navigator of the digital maze, who leverages his data for personal gain while maintaining control over his digital footprint.

Jonathan's story in this new era is not just about the technology he uses but about the transformation of his relationship with his own data. It's a narrative of empowerment, control, and profit, painting a picture of a future where personal data is not just protected but also lucratively harnessed by its rightful owner.

01100010 01110010 01100101 01100001 01110101

After being diagnosed with a terminal illness, a rare neurological disorder, Rose, our octogenarian resident of Seattle, faces her situation with a sense of purpose and resolve. In an act of altruism and in the hope of contributing to medical advancements, she decides to share her medical data for research purposes. Utilizing her PDVK, Rose shares her comprehensive medical data with a broker that specializes in connecting individuals with medical and pharmaceutical research companies. This broker, operating under the principles of C3 and TAPER, ensures that Rose's data is used solely for the purpose she intends: advancing medical research in the field of rare neurological disorders.

Rose's decision to share her data is empowered by the consent and control offered by PDVK. She is able to specify which aspects of her medical history are shared and with whom, ensuring that her personal information is used in a way that aligns with her values and wishes. The broker facilitates this data exchange, connecting Rose's invaluable medical data with researchers who can potentially use this information to make breakthroughs in medical science. This has transformed medicine—the years of scientists struggling to find large cohorts to study and gather enough data about are gone.

This scenario illustrates how individuals have the power to make meaningful contributions through their personal data, with the assurance that their privacy and preferences are respected. Rose's story is an example of how personal medical data, when shared ethically, can have a profound impact. In Rose's case, it's a legacy that transcends her own life, contributing to a greater good.

The story of Rose's decision is a testament to the potential of ethical data management and the profound impact that individuals can have in a world where C3, TAPER, and PDVK are the norm. Her story is another example of how data independence and personal agency in data sharing can transform individual experiences into contributions for the betterment of society.

01100010 01110010 01100101 01100001 01110101

In January 2028, Kara turned eighteen and embarked on a strategic partnership by sharing her PDVK with a data broker whose values deeply resonated with her own. This decision marked the beginning of a unique financial journey, leveraging her digital footprint across various platforms. Kara, with her diverse interests ranging from K-pop to travel, sought not only to share her data but also to ensure it was used ethically, aligning with her belief in responsible data use and individual rights.

Kara's financial acumen was as sharp as her understanding of the digital world. She had meticulously analyzed economic trends over the past fifty years, gleaning that the average inflation rate stood at about 4 percent, while the S&P 500 had returned an average of 7 percent annually. These figures, along with the data brokerage firm's values and mission, formed her decision to enter into an agreement with the broker, who would deposit $100 into her account each month starting in 2028. This amount would then be adjusted annually for inflation at the observed average rate of 4 percent, growing in tandem with an expected yearly return of 7 percent. Kara was intrigued by the concept of compound interest and how it could exponentially increase her savings over time.

Kara understood that this approach to savings, compounded by the historical data on inflation and investment returns, was not merely a testament to the power of financial planning but also an astute use of historical insight for future gain. By the time Kara would consider retirement at the end of 2078, this methodical and strategic approach was projected to transform her initial monthly deposits into a significant sum, illustrating the profound impact of compound interest and the smart adjustment for inflation over fifty years.

This narrative is not just about the intersection of digital identity and financial growth; it's a demonstration of how historical financial insights can be leveraged for future prosperity. Kara's journey high-lights the importance of aligning digital data sharing with ethical practices and financial strategies that mirror one's values and long-term vision. With her broker's arrangement, starting with a $100 monthly deposit in 2028, adjusted annually for inflation at 4 percent, and com-pounding at an expected annual return of 7 percent, Kara's account is projected to grow to more than $12,000,000 over fifty years. This figure embodies the essence of Kara's savvy—her ability to foresee and harness the power of her digital and financial decisions for a secure and prosperous future.

Kara and her peers are now growing up and entering adulthood well aware of their data privacy—and data independence.

01100010 01110010 01100101 01100001 0110101

In a world reshaped by the principles of C3, TAPER, and PDVK, how would Tom's situation unfold? Indeed, Tom's journey through personal challenges unfolds with a sense of empowerment and privacy. Let's walk through his situation in the new data world paradigm.

As Tom seeks online resources to cope with his recent breakup and emotional healing, his digital footprint is now protected by the choices he makes using his PDVK. This means that his searches about overcoming heartache, his engagement with forums on emotional healing, and even his increased interaction with content on mental well-being are under his control. He has the power to decide if and how this data is used by external entities. He made those decisions through his data broker, whom he entrusted with his PDVK.

The influx of targeted ads that once bombarded Tom is no more! He will not receive unsolicited recommendations for herbal remedies like St. John's wort or an overwhelming number of mental health app advertisements. The ads for dating sites, travel destinations for solo travelers, and local singles events are filtered based on the preferences he has set through his PDVK. Tom's social media feeds are no longer automatically populated with articles on self-improvement and promotions for gym memberships, unless he chooses to allow such content.

This new world of C3, TAPER, and PDVK ensures that Tom's quest for personal answers and solace in the digital space remains just that—personal and tailored to his comfort level. Ultimately, though, what he most appreciates is that he can find his solace and healing in *private*. He—like all those in this new data paradigm—understands that without privacy, he has no freedom. The lines between helpful guidance and intrusive marketing are clearly defined by his choices. What was once a bombardment of options and paths has transformed into a curated journey of self-discovery and healing, guided by Tom's control (of own preferences) and consent.

Tom can recall years past, when, in 2019, the FTC sued an online dating service for using deceptive advertising practice; it's evident that such scenarios become increasingly unlikely in this new world. With the strict adherence to the principles of TAPER, companies are held accountable for transparent and ethical uses of data, based on what each individual decides through C3 and PDVK. Deceptive practices like luring consumers into paid subscriptions under false pretenses are curtailed by the accountability and transparency mandated in this new era of digital interaction. Consumers like Tom are now shielded from such exploitative tactics, ensuring a digital environment that respects individual choice and promotes genuine interaction.[71]

---

71 "Match Group, Inc.," 2019, https://www.ftc.gov/legal-library/browse/cases-proceedings/172-3013-match-group-inc#:~:text=The%20Federal%20Trade%20Commission%20has%20sued%20online%20dating,of%20consumers%20into%20purchasing%20paid%20subscriptions%20on%20Match.com.

CHAPTER RECAP

In fact, all these profiles reflect a new digital and data world that is about individual empowerment. The power resides where it should—not with the digital feudal overlords, financial institutions, retailers, etc.—but of the people, by the people, for the people. The power of the people whose data has, until the PDVK, provided untold riches to those collecting it, but not the people themselves.

While the collection of our data may call to mind coupons and ads in our social media feeds, the issue is much darker and urgent than that. Consider this agonizing comment submitted to the FTC in response to its Commercial Surveillance proposed rulemaking:

> "'When I had a miscarriage in 2016, due to my preparations for the baby up until the day the fetus expired, I was plagued with ads for infant supplies for months. [...] Everywhere I looked I was reminded of my loss.' She ends by lamenting her powerlessness in avoiding these distressing targeted ads and the 'advertiser surveillance' that enables them."[72]

Our profiles, as well as each of us, have the right for what goes on in our lives, regardless of what online searches we conduct, what we discuss in front of our phones, what we purchase both

---

72   Arielle S. Garcia, "What do we say to Emily? The human cost of advertising data abuse," 2024, https://www.adexchanger.com/data-driven-thinking/what-do-we-say-to-emily-the-human-cost-of-advertising-data-abuse/.

online and in brick-and-mortar stores, what our credit card and bank statements reflect, to remain private. Our triumphs and our tragedies should be shared when *we* choose, and despite things like coupons for frozen waffles seeming innocuous, it is still the result of data creep and encroaching invasion of privacy.

The PDVK will give our privacy—and our freedom—back to us.

PART III

Data Dreams

and Demands

Future Innovations: Possibilities and Warnings for Big Tech, New Data Ecosystem, and Artificial Intelligence

The development of full artificial intelligence could spell the end of the human race . . .

—STEPHEN HAWKING

The future is here. And we the world's digital citizens have an opportunity. We can enter an era where the digital ether binds us closer than ever before, where the flow of data outpaces the currents of the mightiest rivers—and we can determine the rights and responsibilities we demand for ourselves and the expectations and demands we have for Big Tech and all those who collect our data and invade our privacy.

The principles of consent, control, and currency—our C3— along with transparency, accountability, protection, equality, and respect—our TAPER—can form the bedrock of our times. They are not merely abstract ideals but tangible rights that belong to every citizen of the digital world.

Our digital interactions, guarded by the principles of C3 and TAPER, can foster communities and unearth commonalities. They can allow us to see beyond the pixels and bytes to the human essence that connects us all. In this realm, every click, every share, every moment of digital engagement becomes an act of trust and a step toward mutual understanding.

In our interconnected world, the ripple effects of data breaches, the storms of misinformation, and the tremors of digital division threaten the very fabric of our global society. We see it all around us—online, in our newsfeeds . . . and seeped into our real-life world. Our collective security, our shared prosperity, and, indeed, our peace depend on our ability to navigate these challenges together, guided by the light of C3 and TAPER. This journey toward digital peace requires us to reexamine our attitudes—not just as users or consumers of technology but as active participants in a digital ecosystem that thrives on respect, equality, and shared responsibility. This is a surmountable challenge and one defining mission of our times.

For in the end, it is not just our data that we seek to protect but our dignity, not just our information but our integrity, not just our privacy but our freedoms. In securing these, we secure the future for ourselves and generations yet unborn, a future with possibility and promise.

As we embark on this noble quest, let us remember that in the realm of digital peace, as in all things, we are stronger together. Let our differences not divide us but enrich our collective journey. And let us move forward with the confidence that no algorithm is stronger, no database more enduring, no technology more transformative than the human spirit united in the pursuit of privacy and freedom.

FUTURE AI CONCERNS

In *2001: A Space Odyssey*, the soothing yet vaguely malevolent voice of HAL 9000 echoed through the spaceship. For many viewers of sci-fi movies, they envisioned humanoid or human-sounding robots replacing us. But we're not there yet. The advent of AI has indeed ushered in a new era of technological advancements, yet it also brings forth valid concerns about control, ethics, and the potential for AI to operate beyond our intended boundaries. These concerns are rooted in the very nature of AI systems, their capacity for rapid evolution, and the potential for unforeseen consequences of their autonomous operations.

My own research during my career in the field of AI, particularly the use of GA to encode and evolve populations of algorithms in a Darwinian fashion, demonstrates the remarkable potential of AI to learn and adapt.

By embedding algorithms into genetic frameworks that allow for crossover and mutation, I demonstrated that even with relatively small datasets, AI systems can evolve, learning not just from data but also from other AI systems, fostering the creation of new, more effective algorithms. I developed these algorithms for NASA's JPL in the late 1990s for the purpose of more precise interplanetary navigation.[73, 74]

Adaptation is possible in artificial systems! In other words, AI systems can learn from each other, creating potentially unforeseen and undesired capabilities in new AI systems.

This adaptive capacity of AI, while impressive, underscores the need for robust ethical frameworks and control mechanisms. There is no current unified approach—and in many cases, greed is the driver.

The concept of "data voting," facilitated by the principles of C3 and the use of PDVK, emerges as a pivotal solution in this regard. By empowering individuals to have a say in how their data is used—effectively "voting" with their data—we can ensure that the evolution of AI algorithms aligns with human values and ethical considerations.

---

73   Wassim S. Chaar, Robert H. Bishop, and Joydeep Ghosh, "A Mixture-of-Experts Framework for Adaptive Kalman Filtering", *IEEE Transactions on Systems, Man and Cybernetics* 27, no. 3 (1997): 452–464.

74   Wassim S. Chaar, Robert H. Bishop, and Joydeep Ghosh, "Hierarchical Adaptive Kalman Filtering for Interplanetary Orbit Determination", *IEEE Transactions on Aerospace and Electronics Systems* 34, no. 3 (1998): 883–896.

The integration of C3, TAPER, and PDVK into AI development and data management would mean that individuals can dictate the terms of their data usage. They can choose to share their data with entities that align with their values and visions, making a conscious decision about the role their personal information plays in shaping AI's future. This approach not only addresses privacy concerns but also contributes to the development of AI systems that are more aligned with societal values and needs.

AI's rapid development and adaptability present challenges. My hope is that the conversation we have started means we can steer AI development toward a future where technology not only advances human capabilities but also respects and upholds our collective values and ethical standards.

FUTURE INNOVATION: SOLVING CURRENT CHALLENGES AND ADVANCING HUMAN-CENTRIC VALUES

The integration of AI technology with the principles of C3, TAPER, and PDVK concept offers promising solutions to several current challenges in the digital world and heralds a transformative era. This synergy not only addresses current challenges but also paves the way for a future where digital interactions are secure, ethical, and empowering. Let's explore how these tools can address some of these issues.

First, the PDVK empowers individuals with unprecedented control over their data. It allows people to set preferences for how, where, and by whom their data is used, and about their personal participation in the data currency. This control transforms data from a passive asset into an active tool for personal empowerment, fostering a sense of ownership and autonomy in the digital realm. At last, individuals will have the power.

Next, the PDVK, integrated with AI algorithms, could offer a robust solution to the spread of fake news and doctored images. It could be an essential weapon in the war against misinformation. By seamlessly authenticating each digital interaction and transaction, it ensures the integrity of information. This verification acts as a powerful shield against the tide of misinformation, bolstering public trust and discourse. AI, in fact, guided by the principles of C3 and TAPER, can tailor digital experiences while respecting user privacy. Personalization no longer means intrusive data harvesting but becomes a consensual and transparent process. Users enjoy customized content and services, knowing their data is used ethically and with their express permission.

Another value of the PDVK is that it creates a secure digital identity for each user. This security measure significantly reduces the risk of identity theft, as personal data is locked behind the user's consent. It's akin to having a digital guardian, ensuring that one's online presence is safeguarded against unauthorized use.

By embedding TAPER principles in AI development and deployment, we also demand with our data vote that these systems are designed with respect for human dignity and rights. This approach prevents biases in AI, promotes equal treatment, and ensures that AI serves humanity's broader interests, rather than creating divisive or harmful technologies. The principles of equality and respect within TAPER advocate for inclusive AI and technology policies. This commitment ensures that the benefits of AI and digital advancements are accessible to all, bridging the digital divide and promoting global digital literacy.

By applying the PDVK in *health care*, AI can utilize patient data for better diagnosis and treatment while strictly adhering to consent and privacy norms. Patients control who accesses their medical data and for what purpose, ensuring that their personal health information is used responsibly and beneficially.

AI, guided by the principles of TAPER, can analyze vast datasets related to *climate change*, biodiversity, and pollution, contributing to more effective environmental policies and actions. The PDVK ensures that environmental data collected by entities, often sensitive and critical, is shared and used in a manner that respects public interest.

In the *financial sector*, AI can assist in providing personalized financial services to underbanked populations, following the principles of equality and respect. Simultaneously, the use of PDVK can significantly reduce instances of financial fraud, ensuring that personal financial data is not misused or accessed without explicit consent.

AI can revolutionize *education* by providing personalized learning experiences based on individual learning styles and needs, while adhering to the C3 principles. The PDVK system ensures that educational data is used ethically, respecting the privacy of learners, and contributing to more equitable access to education.

The *advertising* industry, which is undeniably integral to the global economy, faces myriad challenges in the ever-evolving digital landscape. Concerns around data privacy, ad fraud,[75] changing consumer expectations, and the rapid evolution of technology[76] necessitate a transformative approach. The adoption of ethical data practices, encapsulated in the principles of C3, TAPER, and PDVK, offers a pathway toward a more sustainable and consumer-centric advertising future. Implementing the principles of C3 could revolutionize the way consumer data is used in advertising. It ensures that data is used ethically, with explicit consumer consent, that consumers

---

75   Augustine Fou, "Ask them to prove to you that it's not ad fraud," *Forbes*, 2021, https://www.forbes.com/sites/augustinefou/2021/03/31/ ask-them-to-prove-to-you-that-its-not-ad-fraud/?sh=251cd8b67a9d.

76   Maksym Kovalenko, "AI's influence on the programmatic advertising industry," *Forbes*, 2023, https://www.forbes.com/sites/forbestechcouncil/2023/12/05/ ais-influence-on-the-programmatic-advertising-industry/?sh=311d8d855994.

have control over their data, and participate in the aspects of data currency. Transparency and accountability are also critical in addressing issues like ad fraud. Since consumers are voluntarily participating via their PDVK, fraud is addressed. Ad targeting and measurement become a deterministic exercise secured by the PDVK for data lineage.

In this envisioned digital future, every interaction in the digital ether is a step toward a more secure, just, and inclusive world, harnessing the power of data and AI to enhance, not undermine, our humanity. The applications highlight the versatility and potential of integrating AI with the C3, TAPER, and PDVK framework. It shows a future where technology is not only advanced but also aligned with ethical, equitable, and human-centric values, offering solutions across various sectors from health care to environmental protection, finance, education, advertising, and many more.

The shift toward ethical data practices is not without its challenges. It requires significant changes in current business models and practices, investment in new technologies, and a commitment to upholding ethical standards. The paradigm shift I am suggesting has not happened yet. We're all running a five-minute mile.

FDTC OR OTHER AGENCY

I mentioned the FDTC (which does not exist yet!) earlier in the book. The concept of establishing a specific agency to uphold a Data Constitution, including principles like C3, TAPER, and PDVK, raises important considerations. The FTC already plays a significant role in consumer protection and the enforcement of privacy laws. Expanding its mandate to specifically include the enforcement of a Data Constitution could be a feasible approach. Trade in the digital age is driven by massive data. Data and the digital ecosystem are the fuel driving trade today. The FTC's existing infrastructure and expertise in dealing with consumer rights and digital privacy make it a viable candidate for this role.

Alternatively, the creation of a new agency dedicated solely to data rights and digital privacy could offer a more focused approach. Ultimately, though, it is not for me, or Big Tech companies, Big Business, one zillionaire, or one politician to decide any of this. We all need to be part of this discussion. But the FTC becoming explicitly the FDTC would be a natural candidate, in my opinion, as data intertwines heavily with trade and the FTC mission.

CHAPTER RECAP

Considering that AI has the capacity to generate and educate subsequent AI systems, the visions of the future presented by Arthur C. Clarke seem increasingly plausible. However, our current interaction with AI is not without its complications, highlighting the urgent need for a cohesive strategy. AI and other emerging technologies are propelled by data—a critical resource. The solution lies in returning control of this data to its rightful owners: individuals. By empowering people to actively participate and exercise their "data vote," we can ensure they play a significant role in molding the trajectory of AI and future technological advancements. This approach not only addresses present challenges but also secures a more inclusive and democratically shaped technological future.

Fortunately, I think we can come up with a Data Constitution that would engrain these fundamental concepts and truths into the very fabric of our digital society.

CHAPTER 11

Our Data Constitution: On Creating a Future of Freedom

*Political freedom is to be cherished indeed.
But there is no political freedom that is
not indissolubly bound to the inner personal
freedom of the individuals who make up that
nation: no liberty of a nation of conformists,
no free nation made up of robots.*

—ROLLO MAY

The very origins of our nation grew from the nascent seeds of freedom—no taxation without representation, for example. Admittedly, those views of freedom are now, with the backward lens of history, imperfect. Freedom applied to white men only for far too long.

Yet that burning desire for freedom is a clarion call we all seem to cherish—freedom of speech, and assembly, freedom to be who we are, most especially in the privacy of our homes. However, we now understand, through the course of this book, that there is a new aspect of our lives where, despite appearances, we are not as "free" as we thought, and we are being manipulated by the data overlords at a scale unheard of in history.

Yet, the framers and those throughout our history have had a "remedy" when our freedoms were threatened, or when, through the course of enlightenment, it was realized we can and must do better. That was through amendments.

Figure 11.1: Drafting a new kind of Constitution

In the meantime, trying to ensure freedom from the robber barons of the past, the Sherman Antitrust Act was born. What are the modern day's digital overlords but the robber barons of today? The barons of yesteryear used obfuscation, backroom deals, bribes, strongarm tactics, and more—and used our country's rich resources to get ahead in the "currency" of the time, for example, land, coal, and oil.

But as this book has made clear, there is a new currency in the world—it is the new currency of the data economy. And it belongs to each of us, our digital DNA.

As I wrote this book, I pondered, *Why do we not have a cabinet secretary of data and technology?* How are we not recognizing this economic and societal power—and how is it being wielded without our consent? As I mentioned before, I personally came to believe that the FTC, which is an independent agency of the US government, designed to enforce civil antitrust law and to promote consumer protection, is uniquely positioned to own the data enforcement aspects. As an independent agency, the FTC operates outside of the direct control of the president or the executive branch. Yes, the president does appoint the FTC commissioners, with the advice and consent of the Senate, but the agency operates with a degree of autonomy to carry out its legislative mandates. This structure allows the FTC to perform its duties with a level of independence from political pressure, focusing on its dual missions of protecting consumers and promoting competition. Its independence is crucial for its role in regulating business practices and ensuring that markets function efficiently and fairly. Regulating data practices is essential to regulating business practices and ensuring consumers protection, competition, and effective markets. Data is a federal issue and should not be a state issue, as data fuels commerce across the nation. FTC should indeed become the FDTC (Federal Data and Trade Commission).

Thus, I decided to create a Data Constitution. *This is simply a conversation opener.* We need legal minds, privacy activists, those in data, visionaries, AI experts, business leaders, and ordinary citizens to come together and decide the values that represent us in our new data ecosystem.

For now, let us begin with the preamble, which illustrates our commitment to the C3s and TAPER—the fundamental ethics and foundation of how we must approach data.

THE DATA CONSTITUTION: WE THE PEOPLE

The Preamble

In the dynamic and ever-evolving landscape of the digital age, we recognize the foundational role of data as a cornerstone of progress, innovation, and societal well-being. It is as much a piece of our liberty and freedom as our pursuits of happiness. We, the people, the custodians, users, and beneficiaries of this digital universe, in our collective pursuit of a harmonious, secure, transparent, and equitable data ecosystem, hereby establish and affirm this Data Constitution.

Our aims are clear and resolute:

- *To uphold robust and transparent data governance*, ensuring that the stewardship of data is conducted with integrity, accountability, and a commitment to the common good, as outlined in the C3 and TAPER commitment (Article I).

- *To steadfastly protect data rights and privacy*, championing the dignity and autonomy of every individual in the digital realm, safeguarding personal information, and upholding the freedom to control one's digital footprint with transparency (Article II).

- *To fortify data security and protection*, recognizing the vital importance of shielding our digital assets from threats and breaches, thereby ensuring a safe and resilient digital infrastructure for all (Article III).

- *To embrace ethical principles in data usage*, fostering an environment where data is utilized for the advancement of knowledge, societal benefit, and equitable progress while shunning practices that undermine public trust or individual rights (Article IV).

- *To establish a dynamic framework for the evolution of our Data Constitution through a transparent and inclusive amendment process*, enabling us to adapt and grow in harmony with technological advancements and changing societal values (Article V).

- *To declare the supremacy of this constitution in our digital dealings, affirming its role as the guiding document in our shared digital journey,* and pledging our collective adherence to its principles and mandates (Article VI).

- *And finally, to set forth the path for the ratification of this constitution, laying the groundwork for its adoption* and signaling the commencement of a new era in data governance (Article VII).

In this spirit, we unite under this Data Constitution, committing ourselves to a future where data serves as a beacon of innovation, trust, and cooperation, for the betterment of humanity and the world we share.

The Articles of the Data Constitution

I am well aware why we cannot have nice things. Any time we try to do something constructive in the world, there are always those who want to tear down the good or tell us why we can't overcome whatever it is we are attempting.

Funny, perhaps because I have been in a room with scientists at NASA's JPL who were working to land a spacecraft on the red planet of Mars, funding me to research new ways for interplanetary navigation and who, like myself, don't see obstacles as unbeatable. I just think—again, with the mindset of assembling the smartest people in the world for space research—we need to assemble the best and the brightest and begin discussions that address each of the articles we defined in the preamble—but also the details as described here. Again, this will require a broad assembly of experts—and for once, the loudest voices should not be the digital robber barons.

ARTICLES

ARTICLE I—DATA GOVERNANCE

Section 1: Establishment of a Data Governance Body

Section 2: Roles and Responsibilities

Section 3: Data Governance Policies

Section 4: Enforcement Mechanisms

ARTICLE II—DATA RIGHTS AND PRIVACY

Section 1: Individual Data Rights

Section 2: Right to Privacy

Section 3: Right to Data Portability

Section 4: Right to Be Forgotten

ARTICLE III—DATA SECURITY AND PROTECTION

Section 1: Data Security Protocols

Amendments for the Future

As I wrote earlier, the founding fathers crafted an amazing document—but understood it was a living document with mechanisms to address changes. The Data Constitution is no different. It is a planned living document. This is imperative as we could never have predicted all the changes to our lives as a result of data and technology in recent years, and with the advent of AI and other advances, there is no telling what is coming.

Article V of the Constitution provided a method for amendments—a prescient process. Since 1789, twenty-seven amendments to the Constitution have passed. For the Data Constitution, a series of articles and amendments can be added to address specific issues such as international data transfer, AI, and many other topics.

To develop each section would need detailed clauses, reflecting the specific needs, legal requirements, and ethical considerations relevant to the particular data environment it is intended to govern. The book is not intended to be a legal text.

But most importantly, as we explored in chapter 6, we now know that the PDVK could solve so many of the issues the Constitution raises.

The time to declare our data independence is now!

CHAPTER 12

Conclusion...
and a Dream

Injustice anywhere is a threat to justice everywhere.
—MARTIN LUTHER KING JR.

A surveillance state, a world in which we are being watched and "stitched together" 24/7, a world in which our phones, televisions, GPS, and satellites in the sky are spying on us.

The future is now, and . . . *they know who you are.*

In the journey we have taken together in *Data Independence*, what has been the most important aspect of my motivation for writing this book is the *conversation.* I do not think a single intelligent person in the modern digital age is unconcerned about privacy.

When we began on the digital journey—as a global economy, as digital citizens—it was first based on improving our online experiences, based on giving us coupons and ads and internet searches and purchases that were seamless. We logged onto Facebook to see pictures of back-to-school photos of our friends' kids and memes of cats. But we've come a long and very circuitous way since those first coupons, first tracking, and first pictures we uploaded to say "Hello, world. I am here."

Over time, we have allowed those who collect our data to control not only that asset but also the conversation. It is time for the digital citizens of the world to take back and control the conversation. It is time for us to demand our privacy. As we have explored, principles of equality and even the very threads of democracy depend on our data.

Of course, economies of scale depend on our data as well—everything from advertising to consumer goods to health care to hedge funds and insurance and the financial industry uses our data. But the benefits of that relationship—at least economically—generally flow one way.

As I considered these issues, I looked at the foundations of our C3 and TAPER paradigm, our Cs are *consent*, *control*, and *currency*. The C3s developed out of the concept that we the people need to consent to how our data is gathered and used, control that use, and benefit from the data currency that powers the current global economy.

TAPER, as we explored, is:

Transparency (T): Ensuring clear and open practices in data usage, storage, and sharing.

Accountability (A): Entities handling data must be accountable for their actions and compliance with these principles.

Protection (P): Strong measures to safeguard data against unauthorized access and breaches.

Equality (E): Guaranteeing that all individuals have equal rights and opportunities in the digital realm, irrespective of their background or abilities.

Respect (R): Upholding the dignity, privacy, and rights of individuals in all data-related processes.

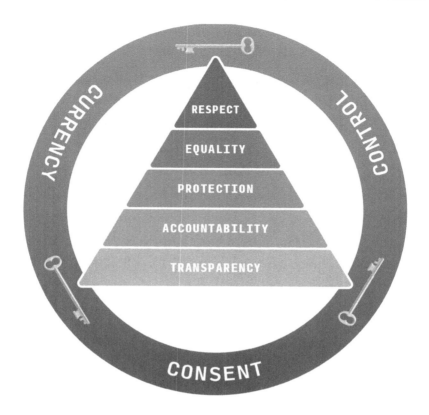

Figure 12.1

Figure 12.1 shows the graphic we used earlier explaining this relationship.

As we explored in the book, one solution I propose to this data quandary is a "personal data vault key"—referred to throughout the text as the PDVK. This tool is a way to, as discussed in chapter 8, address some of the things we all can identify as problematic in our digital universe—whether that is fraud, deep fakes, or the invasions of privacy wrought by the digital surveillance we all face. In addition, it takes back control—one of our C3s—and allows the person who *owns* the data (the digital citizen), regardless of whether in the United States, from where I write, or across the globe in another country, to vote via data brokers for how they wish their digital DNA to be used.

But the PDVK is a suggestion—a conversation starter. What I want this book to accomplish is a frank discussion, with transparency, of what companies, governments, and those who handle our data on a daily basis should be doing to protect our digital DNA. It is also a discussion of who really owns that data—and what sort of financial rewards and incentives are there for the mere click of a button or coupon on an app . . . and who should benefit most.

This was also, at its heart, a book on privacy. We all have heard the adage "that horse has left the barn." Too often, and usually by those with a conflict of interest in this regard, we are told that the issues of privacy and the internet are just too big. They cannot be solved. This is, of course, nonsense. As a nation, as a world, we have tackled grave issues impacting humanity itself—not always perfectly, and not always the way it should be—but we have aimed high. We have had our "moon shots"—whether that has been wanting to land humans on the moon or wanting to tackle the global pandemic of COVID-19 and invent a vaccine in record time.

Data privacy and the issues of data independence are following us well into this twenty-first century, and these issues are only growing exponentially. It is also obvious that the four or five or six (as this number will morph and change) massive, global data companies should not have all the power in regard to our data and the choices made on how to use it.

But I have a dream.

With all honor and homage to Martin Luther King Jr., whom I deeply revere, and in the spirit of the great thinkers who challenged us throughout history to think beyond the state of the world to something better, I propose a dream.

THE DATA DREAM

I stand before you today, a citizen of the digital world, to share a dream that echoes from the technological hubs of the United States to the Cedars of Lebanon, from the bustling streets of India to the vast outback of Australia, from Argentina's Iguazu Falls to the ancient pyramids of Egypt and France with her iconic Eiffel Tower, from the cloud-enshrouded Mount Fuji in Japan to Norway's fjords, from the Philippines' Chocolate Hills to the Great Wall of China, and from *all* corners of our planet Earth.

This is a universal dream, deeply rooted in the global digital dream, transcending borders and oceans, uniting us all in a world interconnected by data. We are each digital citizens of this new world. We increasingly operate in an online realm where our lives and digital DNA are as much a part of us as our "real life."

I have a dream that one day, across this interconnected world, our data—a reflection of our diverse cultures and unique identities—will be proclaimed for its inherent value by its owners, we the people. Defined not by geographical borders but by the universal principles of consent, control, and currency, this value shall unite us all. We, the digital citizens, demand C3. We demand that:

We the people have consent.

We the people have control.

We the people have currency.

I stand before you today, a voice not just of one data scientist or author but of a global digital community, to share a dream that transcends all boundaries and unites us in our shared digital destiny. This dream, empowered by the PDVK and guided by the principles of TAPER—transparency, accountability, protection, equality, and respect—is a universal call to every individual across our interconnected world.

I have a dream that every reader, every global citizen from all corners of the globe, to your family and your neighbor next door, will harness the power of their PDVK. This key is a symbol of unity, bridging our diverse cultures and unique identities. In France, where the spirit of revolution once awakened the world to "liberté, égalité, fraternité," the key will embody the principle of transparency. Across the vast expanses of China, where tradition and innovation merge, it will reinforce accountability. Along the Nile in Egypt, where civilizations were born, it will ensure the protection of our shared digital heritage. In the vibrant heart of Brazil, pulsating with the rhythm of diversity, it will champion equality in every byte of data. And, from every corner of Earth, where respect and honor are ingrained, the PDVK will symbolize the pinnacle of digital respect.

So let privacy reign in every corner of the Earth. Let privacy reign! And when this happens, when we let privacy and freedom reign from every nation and every city, we will speed up that day when all of humanity, regardless of nationality or creed, will be able to join hands—individuals from all corners of the world. We will vote with our PDVK for the values that we support—and we will reap the currency of the value of our digital lives.

I have a dream that from every corner of the world, each individual, regardless of their location, will stand equal and sovereign in their digital presence. This key will not only unlock the potential of data but will also serve as a "data vote," enabling every person to shape the AI-driven future.

This is our hope. This is the faith that I take back to the global digital community. With this faith, and with our PDVKs in hand, we will carve out of the mountain of digital disparity a stone of hope. We will work together, vote together with our data, and shape the future of AI together, knowing that our collective data vote guides us toward a future where technology serves humanity, and not where humanity serves technology.

And on this day, when every individual, empowered by their PDVK and united by the Transparency, accountability, protection, equality, and respect principles, will join hands and sing in the universal language of data freedom and independence.

THE END!